WITHDRAWN
Carnegie Mellon

Perspectives on Sustainable Growth

Series Editor
Min Ding
Smeal College of Business
The Pennsylvania State University
University Park, PA, USA

and

School of Management
Fudan University
Shanghai, China

More information about this series at http://www.springer.com/series/11935

Pierre A. Morgon
Editor

Sustainable Development for the Healthcare Industry

Reprogramming the Healthcare Value Chain

Editor
Pierre A. Morgon
MRGN Advisors
Lausanne, Switzerland

ISSN 2199-8566   ISSN 2199-8574 (electronic)
Perspectives on Sustainable Growth
ISBN 978-3-319-12525-1   ISBN 978-3-319-12526-8 (eBook)
DOI 10.1007/978-3-319-12526-8

Library of Congress Control Number: 2014957479

Springer Cham Heidelberg New York Dordrecht London
© Springer International Publishing Switzerland 2015
This work is subject to copyright. All rights are reserved by the Publisher, whether the whole or part of the material is concerned, specifically the rights of translation, reprinting, reuse of illustrations, recitation, broadcasting, reproduction on microfilms or in any other physical way, and transmission or information storage and retrieval, electronic adaptation, computer software, or by similar or dissimilar methodology now known or hereafter developed.
The use of general descriptive names, registered names, trademarks, service marks, etc. in this publication does not imply, even in the absence of a specific statement, that such names are exempt from the relevant protective laws and regulations and therefore free for general use.
The publisher, the authors and the editors are safe to assume that the advice and information in this book are believed to be true and accurate at the date of publication. Neither the publisher nor the authors or the editors give a warranty, express or implied, with respect to the material contained herein or for any errors or omissions that may have been made.

Printed on acid-free paper

Springer is part of Springer Science+Business Media (www.springer.com)

Contents

1 **Sustainable Development for the Health-Care Industry: Setting the Stage** 1
  Pierre A. Morgon

2 **Can Innovation Still Be the Main Growth Driver of the Pharmaceutical Industry?** 39
  Alexander Schuhmacher

3 **The Importance of Understanding the 'Lived Experience' of Patients in Pharmaceutical Development Programmes** 69
  Kay Fisher

4 **Listening to the Voice of the Patient to Facilitate Earlier Access to Promising Medicines: Interview with Sjaak Vink** 75
  Pierre A. Morgon

5 **Drivers of the Real-World Data Revolution and the Transition to Adaptive Licensing: Interview with Dr. Richard Barker** 81
  Pierre A. Morgon

6 **Sustainable Development for the Healthcare Industry: Vantage Point from Emerging Economies** 85
  Satish Chundru

7 **Disease Management in the Perspective of Sustainable Growth in Health-Care System: Is Disease Management a Good Business Model for the Sustainability of Health-Care System?** 99
  Fereshteh Barei

8 **Thoughts on Sustainable Health Care…in a Patient-Centric Society** 115
  Virgil Simons

9 The Biopharmaceutical Industry is Part of the Solution for Healthier, Wealthier Societies: Interview with Dr. Eduardo Pisani 119
   Pierre A. Morgon

10 Sustainable Development Initiatives: Examples of Successful Programs and Lessons Learned—Interview with Dr. François Bompart 127
   Pierre A. Morgon

11 The Challenges of Sustainable Development for the Health-Care Industry: An Examination from the Perspectives of Biomedical Enterprises 133
   Geoffrey Chun Chen

12 Corporation's Social Responsibility: From the Awareness of Philanthropy to the Demand of Implementation 147
   Vanessa Logerais

About the Editor and Contributors

Fereshteh Barei is an academic researcher at LEGOS Laboratory for Health Economics and Management, at Dauphine University, in Paris. Her research program is focused on Innovation strategy in health-care industry, pharmaceutical and biotech industry, and market access issues. As a lecturer, she discusses strategy, disease management, the case of business model innovation, product differentiation strategies, and evolving R&D systems in international workshops. She is teaching for Master's program in International management at Dauphine University. Since June 2014, she has created Bionowin Santé Avenue Association in Paris focusing on workshop organization, academic research, and consulting.

Professor Richard Barker has been Director of CASMI since its creation and was instrumental in its inception and launch. He is a strategic advisor, speaker and author on healthcare and life sciences.

Richard's 25-year business career has spanned biopharmaceuticals, diagnostics and medical informatics—both in the USA and Europe. Most recently he was Director General of the Association of the British Pharmaceutical Industry, member of the Executive Committee of EFPIA (the European industry association) and Council member of IFPMA (the international equivalent).

His past leadership roles include head of McKinsey's European healthcare practice, General Manager of Healthcare Solutions for IBM and Chief Executive of Chiron Diagnostics. He was also Chairman and Chief Executive of Molecular Staging, a US bioscience company, now part of Qiagen.

In addition to leading and advising a wide range of companies, Richard has advised successive UK governments on healthcare issues, and in particular, on developing, valuing and using new healthcare technologies. His was co-founder of Life Sciences UK, a member of the NHS Stakeholder Forum, and vice-chair of the UK Clinical Trials Collaboration.

He is also Chairman of Stem Cells for Safer Medicines, a public-private partnership developing stem cell technology for predicting the safety profile of new medicines, and a board member of iCo Therapeutics, a Canadian bioscience company.

Richard's book on the future of healthcare; *2030—The Future of Medicine: Avoiding a Medical Meltdown,* is published by Oxford University Press. He speaks frequently on the future of the life sciences, new medical technology and the restructuring of our healthcare systems.

François Bompart, MD is Vice-President, Deputy Head and Medical Director of Sanofi's Access to Medicines department. This department brings together the Sanofi Group's areas of expertise to address the challenge of access to healthcare in developing and emerging countries for specific diseases: malaria, tuberculosis, sleeping sickness, leishmaniases, mental illnesses, and epilepsy.

He also chairs, since 2012, the Global Health Initiative (GHI) of the European Federation of Pharmaceutical Industries and Associations (EFPIA). The GHI gathers the European research-based pharmaceutical companies which are actively engaged in global health. It aims to encourage collaboration among European political actors to create solutions for shared problems in global health. Contact: francois.bompart@sanofi.com

Geoffrey Chun Chen has 7 years of intensive experience in the health-care and biopharmaceutical industry, including 3 years as a Research Manager at Cegedim Strategic Data (currently IMS Health), where he led a quantitative research team to provide data-driven consulting services to Fortune 500 companies. He has also served as a Copresident and Director of sustainable development at Hopkins Biotech Network for 2 years. Currently, he works as a consultant for BioMaryland, a life-sciences-focused department within Maryland's state government. He held a BA in advertising/marketing from Tongji University and an MBA in Health-Care Management from Johns Hopkins University.

Satish Chundru is a seasoned pharma professional and has worked with pharmaceutical companies across countries. He has got a wide exposure to both innovators and generics business. He started his career in digital strategy at Sanofi, France, and in his most recent role was appointed to manage business in Southeast and East Asia for Aurobindo Pharma. Prior to that, he worked in the CEO's office as a junior attaché to the CEO. He is a recipient of Novartis International Biotechnology Leadership Camp (BioCamp) and was invited to Novartis Global HQ in Basel, Switzerland. He holds a Masters degree in business from the Grenoble Graduate School of Business.

Kay Fisher is the founder and chief executive of a leading customer experience company, Experience Engineers. She is a patient-experience specialist and her work has been fundamental in paving the way to driving measurable, improved patientreported outcomes for pharmaceutical companies (Pfizer, Vertex, Boehringer-Ingelheim, Bristol-Myers Squibb) and many other health-care deliverers (Addenbrookes, RNOH, Cancer Care Groups, Bupa). Website: experienceengineers.co.uk

Vanessa Logerais holds a postgraduate diploma in international negotiation (Sorbone Nouvelle), is an expert in corporate strategy and communication, and manages Parangone, a French agency specialized in strategy and communication for sustainable development. She works with public authorities (environmental programs and town and country planning) and with companies (corporate social responsibility

(CSR) policies). Being a journalist and presenter, she regularly hosts national and international events, trainings and conferences, takes part in think tanks and is the author of specialized articles and publications.

Pierre A. Morgon is Chief Executive Officer of AJ Biologics and Regional Partner for Switzerland at Mérieux Développement. He is also Non-Executive Director to the Board of Theradiag since March 2012, a company focusing on in vitro diagnostics in auto-immunity, infectious diseases and allergy, as well as Non-Executive Director to the Board of Eurocine Vaccines since December 2013, a company dedicated to developing nasal vaccines.

He holds a Doctorate of Pharmacy from Lyon University, France, a Master in Business Law from the Lyon Law School and a MBA from ESSEC, France. He is also an alumnus of INSEAD, IMD and MCE executive programs.

Pierre has over 26 years of experience in the pharmaceutical and biological industry, both in marketing positions (from product marketing at country level to global marketing strategy) and in operations (from business unit head to general manager). Through these local and global positions, he has acquired direct experience with blockbuster products in diverse markets (primary care, specialty care, hospital, vaccines, and biotechnology), geographies (US, Europe, Japan, China, India, Emerging Markets) and organizations.

He spent 2 years at ICI-Pharma, followed by 8 years at Synthelabo, then a division of L'Oreal. He joined Aventis Pasteur in 1998, then he had diversified experiences in operations at Yamanouchi Pharma, BMS, Drug Abuse Sciences, Schering-Plough, Bio Alliance Pharma and Sanofi Pasteur. He joined AJ Biologics from Cegedim where he was Chief Marketing Officer and member of the Executive Committee.

Eduardo Pisani As IFPMA Director General, Eduardo leads since 2010 the dialogue between research-based pharmaceutical companies and associations with the United Nations and its specialized agencies. Together with his team and working with IFPMA President and Vice Presidents, Eduardo shares the expertise and perspectives of the industry in global health discussions. He has been instrumental in shaping many of IFPMA's leading initiatives such as the Framework for Action on NCDs and the London Declaration on NTDs.

An industry leader with two decades of pharmaceutical experience, in particular with Bristol-Myers Squibb and Baxter, he has seen health policy and patient access discussions evolve from a close vantage point, and is convinced that platforms for constructive policy discussion are essential to effective policy making. Building on success establishing the first cross-sectoral industry advocacy group on healthcare policy in Brussels, he is similarly engaged in Geneva demonstrating that informed debate and partnerships are important for moving forward. An attorney, Eduardo is Italian and is the proud father of two children.

You can follow him on Linkedin and twitter: @pisanie7

Prof. Dr. Alexander Schuhmacher has been promoted to professor in 2012 at Reutlingen University (Germany). He teaches both in business management and natural sciences, while his research area is innovation management. Prior to his academic career he worked for 14 years in the pharmaceutical industry in various positions in R&D. The most transformative years of his industry career occurred when he was heading the function of strategic planning & business support at Nycomed, whereupon he was also co-heading the R&D integration team following the merger of Nycomed and ALTANA Pharma in 2006/2007. Alexander has in depth experience in project portfolio management, project evaluation, management of pharmaceutical R&D and drug discovery. In addition, he gained general experience in heading and coordinating various senior management committees, as well as in strategic management and change management. Alexander received its Ph.D. in molecular biology at the University of Constance (Germany). He graduated in biology at University of Constance (Germany) and in pharmaceutical medicine at Witten-Herdecke University (Germany) and he is also a graduate of the Executive MBA program at the University of St. Gallen (Switzerland).

Virgil Simons is the Founder and President of The Prostate Net®, a nonprofit patient education and advocacy organization. When he was diagnosed, there was little to no readily accessible information to help prostate cancer patients and their families make informed treatment decisions. Using the experiences gained as a marketing professional, a 19-year survivor of prostate cancer, and a patient advocate, he has built an international organization that uses a matrix of informational techniques (Web site, 800#, email and personal team counselors, public forums, newsletters, and community disease interventions) to address disease risk awareness and early disease interdiction.

Sjaak Vink (Co-founder of myTomorrows and CEO) is a business economist with more than 20 years' experience as an entrepreneur, promoter of innovation and source of inspiration for impact creating entrepreneurs and managing boards worldwide.

He is founder of the Virus of Change and initiator of the Manifesto of Entrepreneurial Change. The entrepreneurial challenge to change the system and improve the lives of many patients around the world inspired Sjaak to co-found myTomorrows.

Chapter 1
Sustainable Development for the Health-Care Industry: Setting the Stage

Pierre A. Morgon

Human Society Has Always Been Focused on Health

Health has been a topic of paramount importance and an integral part of fighting for one's subsistence along with an overlapping with the search for food and shelter, and interestingly it has progressively superseded wealth as a topic of interest. Evidence to that is the Google Ngram chart of the number of books addressing health versus wealth as a core topic; one can only notice that since the 1800s, health-focused books have always outnumbered those related to wealth by anywhere between 10 and 20 % up until the 1920s. From the beginning of the twentieth century, the number of books focused on health skyrocketed (with an acceleration since the 1970s) and the lead of health over wealth as a literature topic is now close to five-fold as expressed in total number of books.[1]

---

[1] https://books.google.com/ngrams/graph?content=health%2C+wealth&year_start=1800&year_end=2000&corpus=15&smoothing=3&share=&direct_url=t1%3B%2Chealth%3B%2Cc0%3B.t1%3B%2Cwealth%3B%2Cc0. Accessed 6 July 2014.

P. A. Morgon (✉)
AJ Biologics, Kuala Lumpur, Malaysia
e-mail: pierre.morgon@wanadoo.fr

Theradiag, Croissy-Beaubourg, France

Eurocine Vaccines, Solna, Sweden

© Springer International Publishing Switzerland 2015
P. A. Morgon (ed.), *Sustainable Development for the Healthcare Industry*,
Perspectives on Sustainable Growth, DOI 10.1007/978-3-319-12526-8_1

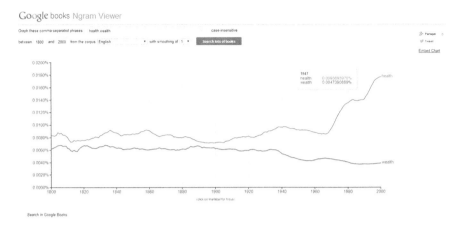

The progressive structuring of human society has led to an organization of more effective care procurement so as to contribute to a healthier society, along with the organization of education and other public services. In this context, the search for medical interventions has progressively evolved from empirical to experimental and scientific, and the life science industry has progressively emerged from the capitalistic efforts to streamline the search for innovative interventions addressing increasingly complex, unmet medical needs.

This evolution has spawned a fully structured industry segment, entirely organized around its capabilities to generate innovation, to protect it with the relevant intellectual property rights, to manufacture its assets in reproducible and high-quality manner, and to commercialize them while complying with a complex set of regulations and guidelines. The industry is nowadays much less driven by its manufacturing capabilities and supply-chain savvy (with the exception of very specific market segments, such as vaccines and some biological products) but rather by its R&D prowess and its superior capabilities to engage with its stakeholders during the latter part of the life cycle of its products, from late stage development to regulatory, then through market access to commercialization.

The Mission of the Health-Care Industry

The health-care industry'ts mission to focus on generating innovative products and solutions, both therapeutic and preventive, for the benefit of the populations around the world, addressing varying types and magnitudes of unmet medical needs, which vary considerably across countries and at times within countries has evolved as pivotal in its strategic roadmap. These geographical differences are not only a source of complexity in the management of the R&D portfolio of the health-care industry and its commercialization policies but also an opportunity to differentiate from competition and to create a form of competitive advantage.

Yet, contrary to the situation in a number of industry segments, the entire value chain of the health-care industry is subject to a large and ever-increasing number of regulations. As an example eloquently described in the guidelines of the International Federation of Pharmaceutical Manufacturers Association (IFPMA), the industry should enforce strict principles of ethical conduct, ensure execution of high standards of manufacturing practices and quality assurance, provide contributions to the overall expertise, and foster collaborative relationships and partnerships with the various stakeholders dedicated to the improvement of public health.[2]

The discoveries of the health-care industry have contributed to changing the face of our world, and the impact has been extremely visible from the demographic perspective, along with the access to clean water and better food, the life expectancy in the developed countries has more than doubled in the last century. This substantial impact on demographics has happened simultaneously to a number of other changes contributing to increasing life expectancy, ranging from better sanitation to access to safer food without supply limitations, and to better living conditions as a whole. As a consequence, the population has been aging and the health-care issues that were linked to the previous societal conditions have progressively given way to more chronic conditions related to aging and to a more sedentary lifestyle. The search for solutions to the ailments linked with modern life in mature economies has evolved accordingly and the R&D efforts focusing on conditions such as hypertension, diabetes, depression, and cancer, to cite only a few, have progressively superseded the search for solutions against infectious diseases in the development programs of the health-care industry.

Simultaneously, a number of initiatives driven either by public authorities or NGOs such as patient associations or international organizations have ensured that the ailments afflicting small patient populations or lower-income regions or countries receive sufficient attention and R&D funding. Nowadays, the majority of leading health-care companies has some form of R&D program dedicated to the neglected conditions and/or "diseases of the South."

It is therefore obvious that the health-care industry's mission is aligned with societal ambitions for a healthier and more sustainable world (Mistra Pharma 2009).

The Life Cycle of Pharmaceutical Products

The health-care industry is working with constrained resources and therefore prioritizing and making R&D choices which are driving the focus of its portfolio. Such arbitrages are not always well understood by the lay public which has sometimes diverging aspirations which it conveys through various media, ranging from the classical tools in democracies, such as voting all the way to the virtual world of social networks. Thus, it is the balance of the perception between these sometimes opposite goals and more specifically the emerging gap between public health objectives and individual expectations, which is increasingly shaping the agenda of the various stakeholders with which the life science industry is interfacing.

---

[2] http://www.ifpma.org/about-ifpma/welcome.html. Accessed 6 July 2014.

Understanding the arbitrages, often referred to as "portfolio management," which the health-care industry has to perform, one has to consider the main characteristics of the life cycle of pharmaceutical products in terms of duration, attrition, and protection.

The development of innovative medical solutions is a long and expensive process, fraught with a high failure rate, in spite of the large number of companies investing in R&D, from the large pharmaceutical conglomerates to the smaller R&D only, biotechnology companies. The development of a novel compound takes, on an average 10–12 years (and sometimes much more), and for several thousands of compounds that are tested during the early development, only a few hundred reach preclinical stage, a handful make it to clinical development, and a few of them reach commercialization stage. Although the probability of success varies according to the type of novel entity, they remain generally low (see chart).[3]

The R&D process of the life science industry has always been marked by high failure rates. From several thousand compounds analyzed during the early phase of exploratory research, a few hundred reach preclinical testing phase, only a handful enter the clinical trials stage, and a small number successfully go through the entire clinical development and regulatory review process. The health-care industry is the largest industry sector in terms of R&D spending (see chart)[4] and yet, its R&D productivity has been declining in recent years despite the so-called biotechnology

---

[3] IFPMA Facts and Figures 2012.
[4] http://www.ifpma.org/fileadmin/content/Publication/2012/IFPMA_CorpBrochureWEBVERSION.pdf. Accessed 6 July 2014.

bubbles.[5] The IFPMA reported in 2010 that the pharmaceutical and biotechnology industries had R&D investments of more than US$ 85 billion (IFPMA 2012) with US$ 48.5 billion R&D investments reported by PhRMA members (PhRMA 2013).

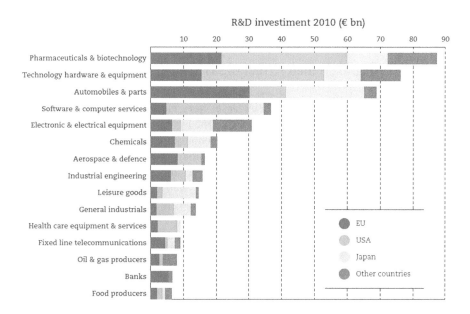

The decline of R&D productivity, both qualitatively and quantitatively, has been extensively commented and attributed to a combination of factors, such as increasing complexity and number of molecular targets, as fundamental science is now more precise; the evolution of the discovery process from random to target based; the larger knowledge gap between fundamental science (proteins, receptors, etc.) and therapeutic applications (hence, the emergence of "translational medicine" as a science); more complex medical needs that need addressing; harder to understand pathological mechanisms of the diseases; increasing difficulties to translate fundamental research into medical discoveries; rising hurdles to identify responder populations; tightened regulation to assess drug safety and more complex evidence demands from authorities; and as a consequence, an increased complexity and cost of clinical development programs, etc. (Paul et al. 2010). Over the years, the focus of R&D has evolved toward specialty care, capitalizing on the progress made by fundamental research in very selected disease areas, with smaller but much better defined target patient populations. Nowadays, the largest number of R&D projects

---

[5] http://www.fda.gov/AboutFDA/WhatWeDo/History/ProductRegulation/SummaryofNDAApprovalsReceipts1938tothepresent/default.htm. Accessed 6 July 2014.

is found in the field of oncology, while the cardiovascular segment is getting much less attention and central nervous system diseases even less so.[6]

Further, the industry is adamant to protect its rare assets as tightly and effectively as possible, in order to balance the staggering R&D costs with a commercialization period under some form of exclusivity.

The Mega Trends Which Affect the Health-Care Industry

During the past decades, health-care expenses have increased steadily despite numerous attempts to curb or limit their growth. The health-care spending is currently representing a very large share of the real economy expressed in terms of GDP. In most countries of the Organization for Economic Cooperation and Development (OECD), this share is growing at such a pace and up to such a level that it is challenging the sustainability of the health-care systems (World Economic Forum 2013). With the exception of the USA where health-care expenses represent a staggering 18 % of GDP, most developed countries are investing 10 % or more of their real economy in health care.[7] In emerging economies, the cost of health care is driving difficult trade-offs: When faced simultaneously with the emergence of diseases which are frequent in mature economies (such as hypertension, diabetes, dyslipidemia, depression, or cancer) while still dealing with remaining diseases typical of the emerging world (diarrheal or respiratory infectious diseases such as cholera, typhoid fever, rotavirus diarrhea, or tuberculosis), the public health decision makers are driven to tough choices to allocate their health-care priorities and resources. In addition, these emerging countries are often facing infrastructure challenges to invest in both the management of the wide scope of medical conditions, and in the construction of a robust, efficient, and accessible primary care infrastructure, including the brick-and-mortar components of hospitals and dispensaries and the human component of properly trained health-care providers.

While the emerging countries are facing these difficult investment choices, the mature economies, which are crippled by their economic debt also face tough trade-offs to drive the mutation of the health-care infrastructure and deal with the public expectation for efficient and personalized care procurement.

In this context, the health-care industry is often considered as one of the key drivers of the progression of the health-care costs, although the share of the total health-care costs represented by medications and devices in the total amount of health-care expenditures is usually comprised between 15 and 20 % depending on the countries. The question therefore becomes one of understanding where the actual drivers of the bulk of the health-care expenditures are and what could be done to manage those effectively.

The health-care spending keeps increasing as it is driven by the "cost disease" as described by William J Baumol (2012), as health-care procurement is a very labor-intensive activity.

---

[6] Scrip Research and Pharmaprojects 2013 Citeline.

[7] http://stats.oecd.org/Index.aspx?DataSetCode=SHA#. Accessed 6 July 2014.

The key tenet of the book is that all services that consume personnel are condemned to see their cost rise at a rate significantly greater than the economy's rate of inflation because the quantity of skilled labor required to produce these services is difficult to reduce. It is especially the case in health care as automation or standardization is not always possible, and also since labor-saving productivity improvements occur at a rate well below the average for the economy (due to the increasing use of skilled labor, from an ever-increasing number of health-care providers, always more skilled and specialized). The authors underscore that the enduring stagnancy of the productivity has imposed a cost history of constant rise to the corresponding services. Every patient has to be examined individually; the patient is subject to specific examinations, possibly by a different health-care provider of different specialties until the diagnosis is firmly established, the course of treatment is decided, and a successful outcome is reached. Therefore, at every step of the patient's journey, there is the intervention of one or more health-care provider whose role, due to the evolution of science and the better understanding of the disease and their treatments, is ever more specialized. These well-trained, high-quality resources are therefore very expensive. The honorarium of health-care staff is subject to increases like salaries without much productivity gains, since by definition a diagnosis is personalized to each patient case. One could argue that automation and productivity gains are happening in other quadrants of health-care procurement chain such as biological or radiological examinations, yet the benefits of such gains are more than offset by the increasing sophistication and number of said examinations.

The authors also point that a simple way to slow down the increase of the costs is to shift some of the labor from the supplier to the consumer. This is already visible in developed countries, e.g., increase in the share of the health-care costs borne by the patients or increasing accountability of the patients in the execution of prevention measures.

They point to the absolute necessity for the high-tech firms to keep investing in the innovation that drives productivity growth. We can observe such initiatives all along the R&D and manufacturing value chain in the health-care industry. They argue that the rising cost of innovation can lead to fiscal strategies "that reduce investment in R&D, thereby further impairing R&D productivity. This in turn can impede overall productivity growth."

They are also arguing against cost controls (even though these are very often the staple of the public health policy in most economies) as such controls lead to the deterioration of the quality of the services and possibly worse to their partial or total disappearance. They are explicitly stating that "crude attempts at budget reductions or price controls in health spending are unlikely to be effective, equitable, or efficient." On the evaluation of health-care interventions (and these include the use of medications, devices, etc.), the authors argue that measuring them in terms of quality-adjusted productivity gains is not appropriate, and that the adequate criteria would be to measure labor-saving or cost-saving enhancements of productivity. In other words, the ability to induce a change in the way care is delivered.

Lastly, regarding global governance and equitable access to care (such as investing in emerging countries), the authors argue that "foreign aid that simply buys health products and services in developing countries might crowd out local government health spending and distort national priorities," hence calling for a different approach to expanding access to care.

The trend of the share that health-care costs represent in the real economy is worrying for the public authorities because the growth in health spending has always outpaced that of the GDP, especially during economy downturns. What is even more worrying is the absence of correlation between health-care expenses and quality of care[8] and that has drawn the authorities' attention on the efficiency of care procurement. It has also been attracting the lay public's attention on the broad—and not always well-defined—topic of the value for the individual. This lack of correlation between the magnitude of investments and the quality of care procurement is also stretching to the relationship between the health expenditure and the degree of satisfaction that consumers have with their health-care systems.

The absence of a solid correlation between investments in health care and quality of care is triggering a host of activities driven by the authorities across all countries to search for efficiencies, looking for benchmarks from abroad to identify more effective ways to utilize health-care resources. The authorities are also concerned by the poor correlation between health-care spending and patient satisfaction[9], as the topic is politically loaded and oftentimes central to electoral programs. From the patient's perspective, the increasingly ubiquitous availability of information and the behavioral evolution toward challenging both the medical and the public authority are fueling mounting patient-driven challenges.

In addition, the life science industry has to cope with other global macro trends that affect its relationship with its ecosystem and stakeholders.

First, the shift of economic power to emerging markets is driving global economy growth and the rise of other health-care expectations while the developed countries are struggling with debt. Therefore, in both cases, social, political, and economy pressures result in investments in health-care infrastructure and increasing price pressure on medical interventions. The mature economies are looking for more cost-effective ways to utilize their health-care resources without raising public concern over care rationing. The emerging economies have to arbitrate between health-care priorities and invest in infrastructure to ensure the proper level of care coverage, access, and quality. In the first instance, the health-care industry has to find the right balance between innovation and the price it charges for the said innovation so as to generate an economic and social surplus for the health-care system. In the second instance, the industry has to balance its portfolio to ensure its adequacy with the local epidemiology and public health priorities.

Second, the emerging markets will represent the largest share of the growing cohort of health-care consumers, thanks to the swelling of the ranks of the middle class, driven by the economic growth. Simultaneously, these numerous and eager health-care consumers will increasingly live in urban areas; as a consequence, they will have easier access to care procurement as well as to health information, leading to an increase in qualitative and quantitative demand and in the challenge of the medical and public authorities. The flipside of the increase of the urban popu-

---

[8] Euro Health Consumer Index Report 2009.
[9] Analysis of data from WHO, The Commonwealth Fund, Frontier Centre.

lation will also take the form of a less-active lifestyle and evolution of the dietary regimens, resulting in an increase in incidence of the chronic diseases linked to sedentary behaviors as well as to a greater exposure to poorer air quality, which will trigger a higher incidence of respiratory conditions.

Lastly, whether in developed countries or increasingly in emerging ones, we are looking at shifting demographics with aging populations (also a consequence of the previous trends, better economic status contributing to aging) which will increase health-care demands in relation to the growing incidence of chronic diseases and organ failures linked to ageing.

The health-care industry has always been paying close attention to those trends and they contributed largely to the R&D orientations as well as to the definition of business development and geographical expansion priorities.

Societal Expectations for Personalized Medicine

Besides these economic and demographic trends, one has also to consider the impact of societal evolutions.

In recent years, the combination of education and access to information and mobility has contributed to the growing sense of self over that of the community. While such a trend can be welcomed from the perspective of consumer goods, as it creates opportunities for market segmentation and creation of value propositions matching more closely to the expectations of the customers, it comes with a number of hard to manage consequences in the field of health care.

For instance, when it comes to immunization, we have been witnessing the progressive evolution of resistance against immunization recommendations. Case in point is the growing number of parents who refuse to immunize their children against infectious diseases because they do not perceive those as a threat while they are worried about potential adverse events—some immediate, but most importantly the deferred, longer term ones—linked to vaccine administration. Vaccines are meant to protect against infectious, transmissible diseases; it is therefore of the utmost importance for the public health authority to ensure an adequate coverage in order to block transmission of said diseases. Typically, coverage of 80 % or more of any population is considered as an adequate threshold to block the transmission of diseases such as influenza. Because it is so important to reach such high coverage rates of the target population, all the stakeholders involved in executing immunization campaigns are striving to ensure adherence to the immunization calendars or guidelines. Thus, the goal is to mobilize a large enough share of the population to undertake a medical intervention on an otherwise healthy group of subjects, using a standardized medical intervention. The resistance to the recommendations is often stemming from the fear of safety issues as much as from a less precisely defined concern over such "one-size-fits-all" medical interventions. In fact, in an era where patients are expecting customized solutions, the very standardized nature of the immunization recommendations is negatively striking the chord of individualism, resulting in skepticism and sometimes resistance.

The growing magnitude of this societal trend has consequences for the stakeholders involved in the design and in the execution of health-care policies. For all the stakeholders, the ever more assertive attitude of patients has been considered very seriously and patient centricity has taken center stage in the past decade. Almost every life science company is claiming in its mission statements that the patient is central to its strategy and driving the focus of its strategic activities, from the priorities of the R&D investment, to the quality of manufacturing, and to the content and quality of the information flows in the context of the commercialization of its products. Such statements related to patient centricity are now so ubiquitous that they are hardly differentiating and it is important to look beneath the surface of the glossy corporate brochures and press releases to assess the reality of patient centricity, and how it translates in the real life of these corporations.

Patients are first and foremost people, living in an information age, accelerated by mobility, in which data are ubiquitous, where access to media is easy, and yet where understanding of available information is leaving room for improvement and is extremely uneven across population subgroups depending on the level of education and command of health-care matters.

It is commonplace to say that what happens in an Internet minute is staggering[10]. People access a deluge of data (including that related to health care) under various forms and from a multitude of sources, accelerated by the connectivity between devices.

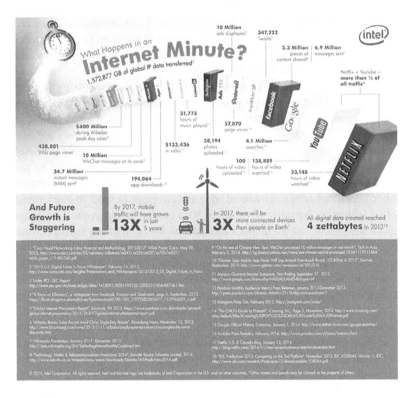

---

[10] https://image-store.slidesharecdn.com/561a0856-b834-11e3-a614-22000a9780da-original.jpeg. Accessed 1 June 2014.

In this context, not one stakeholder can master the full scope of the information flows to build a comprehensive, reliable, and accurate representation of what is broadcasted on a given health-care topic. Subsequently, taking into consideration the voice of the customer is commensurately difficult, in light of conflicting priorities as described below, and compounded by expanding expectations for more effective and better-profiled care procurement. Overall, the difficulty is simply finding what the real voice of the customer is.[11]

As new generations reach adulthood and working age, customers of the health-care system as well as health-care providers are increasingly "born digital"; Internet, e-books, and social networks are all part of the standard tools that they are familiar with and which are an integral part of their daily functioning[12]. As the generations that are defined as "digital natives" grow up and reach the level of influence, starting from voting age all the way to positions of authority, one is witnessing drastic changes in information management practices and we are likely to see a parallel evolution of the way public health priorities are determined and addressed.

When looking at the number of articles published on health-care topics per year from 1970 to 2010, one notices a steady increase of the number of publications over 40 years, with the rate of increase becoming more pronounced from the beginning of the 2000s.[13] This surge in the number of publications is obviously the combined result of intensive research efforts and of the sharing of scientific knowledge. The sheer volume of information that is created is triggering a growing challenge for the human cognitive capacity, due to the exponential increase of both the number of sources and of their complexity, when these are needed for quality clinical decision making.

As the diagnosis of patients and the decision regarding the best course of action to follow are both becoming more complex, so is the task of the various stakeholders aiming at providing reliable, trustworthy, and most importantly understandable information, both to the practicing medical community and to the lay public. Evidently, the same difficulty is amplified downstream at the patient level, since the patient facing health-care providers often do not have enough time or the adequate educational skills to communicate effectively with the patients, most of the latter lacking the relevant educational background to approach data in a discerning manner and to form a relevant and scientifically sound opinion on their own.

The DNA of the Health-Care Industry: Evolving Management of Innovation

The concern for the life science industry is that the return on its R&D investment has eroded despite sustained spending in terms of percentage of total revenues (and the nominal amount of said R&D spend had increased, commensurately to the turn-

---

[11] Economist Intelligence Unit Survey, July 2012.

[12] http://mashable.com/2013/12/21/technology-age-comic/. Accessed 1 June 2014.

[13] Online searches at PubMed http://www.ncbi.nlm.nih.gov/pubmed. Accessed 1 June 2014.

over of the health-care industry) reflecting the lower outputs in new molecular entities, as well as the more limited volume and value sales for each new entity which addresses more targeted patients' populations. Yet the industry maintains its R&D investment, as the analysis of the evolution of approvals by regulatory agencies (FDA and EMA) indicate that innovation is driving sales, albeit at a lower rate of return. It is expected that this trend will continue, as indicated by a pool of evidence ranging from the declarations of the CEOs of the health-care companies, as well as by the sustained flow of investment by venture capitalists and capital developers in innovation-focused start-ups.

The maintenance of the R&D investment should not hide the substantial shift in the mix of projects, in terms of disease areas, and precisely defined patient populations. Even the latest development candidates addressing widespread conditions such as dyslipidemia are subject to a clinical development, resting on a larger number of clinical trials, each of them addressing smaller, well-targeted, and defined patient sub-populations, rather than larger scale clinical studies including less-precisely profiled patients.[14] Evidently, the nature and structure of such clinical studies is driven by the evolution of science as well as of the request for specific types of supportive evidence from the authorities. As the drug candidates address chronic diseases in the management of which the patient attitude is a contributing factor to the treatment outcome, the design of the clinical trials is starting to take this dimension into consideration.[15]

The narrowing of the patient populations initiated by the more precise, science-based profiling is further enhanced by the behavioral dimensions. As the societal trends have evolved and care procurement has grown more customized, these emerging individualized approaches lead patients to expect a fully personalized approach, fueling further the search for information.

What personalized—or "customized"—medicine is should be better defined, as well as the expectations of the patients, and the consequences for the industry. This author has been already writing on this topic[16]. Could medicine be customized, like the newer generations of cars, fully tuned to the customer liking (color, decoration, and features), the women's bracelets with a unique selection of charms, the decoration of our homes, the attention that we get when we meet our private banker of when we flash any frequent user membership card, etc.? In actuality, patients increasingly want medicine to be fully tailored to one individual, and in many instances, they are ready to foot the bill (at least partially) to an extent commensurate to the level of personalized attention that they get. It will probably be analyzed over time if this is a consequence of the societal evolutions or of the technological leaps that are pushing ever further the boundaries of the understanding of the pathologi-

---

[14] http://www.scripintelligence.com/home/ACC-PREVIEW-What-drug-trials-will-be-hot-in-Washington-DC-350858. Accessed 19 July 2014.

[15] http://www.scripintelligence.com/home/Healthcare-2030-facing-up-to-a-pharma-future-346693. Accessed 17 July 2014.

[16] http://brainfoodtv.com/personalized-medicine-who-needs-it-and-what-for/#.U8qQ6LHmcSQ. Accessed 15 July 2014.

cal mechanisms underlying the medical scourges that affect human beings, down to details specific to an individual.

Beyond the expectations of the individuals, the reality of the procurement of care is still a far cry from truly personalized medicine. In the earlier article referenced above, this author has argued that there's a substantial gap linked to the false belief prompted by the buzz on personalized medicine which is falsely leading people to believe that we are seeing already personalized care procurement, whereas we are actually looking only at increasingly precisely profiled medicine, yet still far from a genuinely individualized medicine.

Within such an evolving societal context, the health-care industry has to evolve so as to ensure the sustainability of its innovation-based model.

Sustainability and Pharmaceutical Products: Role in Human Health (Mistra Pharma 2009)

As characterized by several authors, a sustainable society is managing economic, environmental, and social issues in a long-term sustainable way. More specifically, a sustainable society must have a health-care system that is resting on similar principles, including the use of pharmaceutical products. This implies that the entire life cycle, from development and manufacturing to consumption and disposal of pharmaceuticals must be sustainable. These requirements for sustainability apply, whether the health-care solutions are produced locally or imported, as the responsibility for using sustainable products applies globally, irrespective of national regulations or borders (Wennmalm et al. 2010).

For instance, in his book entitled The Soul of Capitalism (Greider 2003), William Greider argues that goods-producing activities that generate increased economic output do not necessarily generate what the society wants and needs. The impact being that social trust is among the casualties of work when ownership is distanced and depersonalized from its real-world meanings and therefore insulated from the real-world consequences. Overall, he pleads for trustworthy financial firms, accountable business organizations, new patterns of ownership and governance, and new mediating institutions.

The author encourages an ethic of shared responsibility between consumer and producer as he argues (convincingly) that corporate governance is a central variable in the ecological crisis. He recommends that corporations develop the capacity and culture to tell the truth (yet without linking this capacity with value creation, or some other means to incentivize corporations and their shareholders to embrace the concept). He argues that effective corporate governance recognizes that motives of self-interest and social obligation are compatible and mutually reinforcing (here again, the link with value created and thus shareholder opinion/orientation remains to be demonstrated).

The notion of individual versus collective benefit surfaces through a discussion on the issue raised by corporate privileges damaging the interests of individual persons. This notion is well understood by health-care companies, as it is the basis of most "health technology assessment" discussions with the authorities.

The author argues that a "new social corporation" should have the following founding principles:

- Producing real new wealth
- Achieving harmony with nature (apparently defined as the corporation ecosystem)
- Having governance mechanisms to ensure participatory decision making and equitable adjudication of inevitable differences (note: some legal dispositions in Europe about company profit sharing schemes are already in existence)
- Undertaking concrete covenants with the communities that also support it
- Promoting unbounded horizons for every individual within it (hence all the regulations on employee training, career advancement, etc., yet used in a discriminative way for obvious reasons)
- Designing a culture that encourages altruism
- Committing to defending the bedrock institutions of the society (in that case, the US, hence defined as viability of family life, integrity of representative democracy, etc.).

The author makes a brief set of comments on the pharmaceutical industry, blaming it for riding free on the public funding (NIH research) to develop patented medicines and deriving profits from this IP protection through inflated prices (note: it is accurate that drug prices in the USA are the highest in the world). The point would gain in accuracy if it was distinguishing between investments in research versus in development and acknowledging the development risk and costs, as well as the recent regulations encouraging generics, reference pricing, the use of health economics to measure the cost-effectiveness of drugs and determine their prices and reimbursement, etc. Also, some of the author's suggestions have since been put into legislation (e.g., Sunshine Act, protection of whistleblowers, etc.).

An interesting question is that should the corporations be more farsighted in focusing on what society wants and needs for its distant future, it would require a more precise definition of these needs and wants, and it would require to define these as goals for corporations, among other financial (and more short-term) goals. The "sustainable development" question is "are these reconcilable?" The author argues at several points that top-down change is not possible (i.e., legislation-driven) but that grassroots approaches (experiments undertaken by entrepreneurs) are mandatory, subsequently disseminated via compelling stories. Consistent with the notion of "stories", he mentions "indicators" which "work in two relatively invisible dimensions: individual consciousness and social process." And this analogy to "viral change" has also to be put in perspective with the recommendation for mediating institutions, likely assuming that corporations will not be accountable right away and will need time to regain trustworthiness.

The process of using social indicators to construct social narratives applies the principles of ecology to human systems (economy and community), to suggest solutions: what has to be altered to restore the balance to the ecosystem (a form of homeostasis) and, thus, its sustainability.

Pharmaceutical Products and Contribution to Global Health: Global Health Requires More than Better Drugs

Since the beginning of the industrial age of the health-care industry, the level of consciousness to the sustainability issue has been evolving progressively. Evidence to that is the series of "Pharma Future" reports which have been addressing various aspects of the matter over the past 10 years.

The first such report, "Pharma Futures 1," was issued in 2004 to present the conclusions of a scenario planning exercise executed by the industry and its investors. The exercise was stemming from the early conclusions that the business model of the health-care industry had to evolve in light of the mounting challenges, both internal (such as the R&D drought) and external (such as the financial constraints weighting on most health-care systems; Pharma Futures 1 2004). The report identified an imbalance between the short-term shareholder perspective and the long-term value of research for the other industry stakeholders and highlighted seven key findings. First, the impact of the emerging markets was deemed underestimated, and this proved a very accurate prediction as these countries are now central to the development strategy of the industry. Second, the sustainability of the industry value was described as strongly correlated to issuing innovative therapies, and this finding has also been reinforced by recent analyses. Third, the authors estimated that the industry was in need for a more "adaptive, flexible, and open minded leadership…to signal to the investors the need to change; which has been echoed in several publications, including from this author. Fourth, the ability to change successfully was attributed to seizing the "first mover advantage," yet history has demonstrated that the first mover were often the industry stakeholders, starting by the regulatory authorities. Fifth, the perennial issue of trust was underscored, albeit from the perspective of the investors' confidence that the industry can deliver sustainable shareholder value. Sixth and linked to the first finding, the authors doubted that market-based solutions will help meet the access needs in poorly developed countries; the evolution of specific industry policies has demonstrated that much work remains to be done, but that some market-based solutions such as tiered pricing are indeed proving extremely effective to solve affordability and access issues. Lastly, the growing power of the health-care consumer was highlighted as a driver of awareness and of increased transparency.

The second report addressed a number of critical questions pertaining to "market access," including the mutations of the landscape of payers and also the access to emerging countries, and how the industry could manage its productivity more effectively. This scenario planning exercise is focused on visioning the environment and the health-care ecosystem in which the products under development at the time of the analysis will be launched. The analysis encompassed the projection of the trends as well as their societal consequences and went on to elaborate further on how the industry and its shareholders should address them in a balanced way that would preserve the expectations of financial return of the investors while meeting the expectations of society. As a consequence, the initiative also focused on the need for the industry to communicate clearly and transparently on its strategies to manage these challenges to investors, as well as how investors should signal what needs to be brought to their attention (Pharma Futures 2 2007).

The report reinforced the earlier conclusions of the importance of R&D, when it was already obvious that the technological advances and the sustained investment were not necessarily translating into greater output of novel therapies. It also underscored the need for the industry to respond efficiently to increasing demands for evidence supporting pricing and reimbursement, therefore supporting the value proposition for products addressing predominantly chronic diseases. And the report echoed again the opportunity for growth embedded in emerging economies, provided the industry can address specific public health needs in an affordable manner, working in close partnership with the governments and the civil society.

Unlocking these opportunities was described as raising several challenges, most of which are still current, 7 years after the report has been published. The industry was expected to transition its portfolio to more targeted products requiring an extensive collaboration with a broader range of stakeholders from development to commercialization, and this prediction has proven extremely accurate. As a consequence, R&D was anticipated to be reorganized to eliminate redundancies and ensure a balanced portfolio in terms of types of innovation and probability of success of the programs, including unmet medical needs with a limited commercial potential. Meanwhile, the industry was expected to take a balanced approach on pricing of new products, hence managing successfully to navigate the payers' willingness to pay while ensuring that it is perceived as a trusted partner, hence demonstrating that it adds value to the health-care system. Thus, the industry should be able to contribute to implementing policies through collecting data throughout the entire product life cycle from the perspective of health outcomes as well as health-care system efficiency and societal expectations. In emerging economies, the challenge will include adequate pricing—and once again, this has been successfully achieved through policies such as tiered-pricing—while preventing negative repercussions on more affluent markets and maintaining an open dialogue with local stakeholders to secure the relevant foundational level of trust.

The third report is focused on a pool of opportunity that the industry has been contemplating for its contribution to economic and population growth, namely the middle-income countries including China, India, and Brazil. The purpose of the initiative was to analyze the connections between the public health objective of improving outcomes in these countries, and sustainable pharmaceutical business models. The core topic was affordability, and how the industry should balance its profitability objectives with the markets' willingness and the ability to pay for health-care interventions. It was recognized that the two should be aligned for the industry to perform in a sustainable manner (Pharma Futures 3 2009).

These emerging markets were described as complex from the cultural, ethnic, and economic perspectives, and presenting infrastructure and urbanization challenges; nowadays, these drivers of complexity remain the same, albeit under evolving proportions, hence requiring specific skills and investments. Although affordability was described as a key issue, time has shown that economic growth has spawned the swelling of the middle class which can afford more expensive care, and tilted the gradient of affordability-driven access in the right direction. Lack of proper infrastructure, inadequate access to care in the rural areas, and complex relationships with the local governments often mean that emerging markets are not entirely accessible by the international health-care companies. The topic of the social contract was raised and especially the expectation that life-saving drugs would be made available to all patients who need them, meaning that the privately held industry would be asked to solve the issues of the public health sector through developing innovative business models and ensuring that its shareholders understand the long-term value embedded in such initiatives.

More specifically, it was identified that investors need improved visibility on the opportunities in emerging markets and the expected return on invested resources with agreed-upon, forward-looking performance indicators. Meanwhile, healthcare companies need a greater flexibility in their infrastructure and approaches to pricing and distribution, up to the development of hybrid solutions supported by financing vehicles combining philanthropic and mainstream venture capital. The industry has to ensure that its products meet actual health-care needs in an affordable and accessible manner. Overall, the conclusions advocated enhanced communication as a foundation for accountability and transparency.

Nowadays, most companies have a clear "north–south" policy including a pricing component, often referred to as "tiered pricing," by which the price of a medication is adjusted to the economic status of the countries. This question of affordability is not anymore confined into emerging countries, as indicated by the recent rows over the price of oncology products in the USA[17,18] or over the price of hepatitis C medications both in the USA and in Europe[19,20,21].

Elaborating further on the question of affordability, the fourth report is focused on "shared value," defined as the need for the health-care industry to rebuild its social contract with society at large. The report elaborated further on the shrinking R&D output and the difficulty to meet all patient needs in a context of overburdening debt that reduces the willingness and ability to pay for innovation, which is

---

[17] http://www.nytimes.com/2012/11/09/business/sanofi-halves-price-of-drug-after-sloan-kettering-balks-at-paying-it.html?_r=0. Accessed 15 Aug 2014.

[18] http://www.cancerletter.com/articles/20130628_2. Accessed 15 Aug 2014.

[19] http://www.nytimes.com/2014/08/03/upshot/is-a-1000-pill-really-too-much.html?abt=0002&abg=0. Accessed 15 Aug 2014.

[20] http://www.webmd.com/hepatitis/news/20140714/high-cost-hepatitis-c-drug-sovaldi-investigated. Accessed 15 Aug 2014.

[21] http://www.techtimes.com/articles/12045/20140805/sovaldi-hepatitis-c-drug-at-84–000-per-treatment-course-sparks-healthcare-concerns.htm. Accessed 15 Aug 2014.

a major concern for shareholders focused on the return on their investment. This central question is still burning today and is likely to remain so for the foreseeable future (Pharma Futures 4 2011).

The report highlighted the delicate R&D balance to be achieved between addressing chronic conditions and finding new solutions against resistant infectious diseases and underscored the unique capabilities possessed by the industry to translate fundamental science into approvable products, building on its network of relationships and on its vast knowledge of the diseases. The R&D challenge was—once again—described as one of efficiency since the R&D output kept declining since earlier reports. The attrition was correlated to five drivers: industrialization, duplication, risk aversion, consolidation, and regulatory requirements. As the industry adopted industrial techniques to screen and develop drug candidates, it biased its skills pool away from pharmacology and physiology and lost proximity with integrated biology and experimental medicine. Also, the generalization of the industrial techniques triggered a focus of the industry on the same few leads, resulting in numerous "me too's" and lots of areas of unexplored medical needs, especially for conditions lacking good biomarkers. As the shareholders are keeping a close eye on the return on their investment, the industry grew a tendency to focus more on validated targets, opting for predictable returns, further reinforcing the focus on the same targets and shrinking the breadth of the R&D portfolios. This was further accelerated by the wave of mergers and acquisitions, leading to a global decline in R&D productivity. Finally, the ever more stringent regulatory demands, fueling an increase of the cost of the clinical development and of the risk of failure, drove the industry to make difficult choices to prioritize their lead candidates; and the ensuing rationalization is further fueling the investors' concerns that the R&D strategies may not yield the expected return.

The question of R&D productivity is connected to that of return on R&D investment through the broad question of the value of innovation, for which the surrogate marker is the payers' willingness to pay. As the payers are operating under growing financial constraints and are increasingly concentrated and using more sophisticated methodologies to track outcomes, the industry has faced commensurately increasing resistance to pay for marginal innovation and had to polish her health economic capabilities to substantiate genuine value propositions, reinforcing the decision-making role of the payer over that of the prescriber. The business model of the industry has evolved to place greater emphasis on evidence of value, further fueling the industry's tendency for risk aversion as it has to collaborate with an expanding scope of stakeholders, including better informed and more demanding patient groups. As the report puts it, *"This could evolve into a 'Shared Value' model in which the social contract is renewed to the mutual benefit of industry and society"*.

The report highlighted a number of recommendations, for each type of stakeholder.

Government agencies are expected to provide a clear sense of direction through a health-care strategy and to create the conditions for discussions about future health policies. They should highlight areas of unmet medical needs and facilitate R&D

portfolio orientation choices through enhanced multilateral collaborations, and consistent health, regulatory, and reimbursement policies.

The regulators, including agencies tasked with performing the assessment of novel health-care solutions, and payers are expected to collaborate and harmonize assessment criteria, engage in a dialogue with the other stakeholders to provide guidance, and explore alternative mechanisms of pricing and reimbursement. The dialogue with the industry is especially critical to ensure that all patient needs are addressed, irrespective of the magnitude of the commercial opportunity.

The health-care industry has a commensurate and correlated set of obligations, revolving around the revamping of its business model, the redefinition of its core competencies, the clarity of its strategic roadmap towards investors, the accuracy, transparency, and reliability of its communication and overall the need to be bolder in its search for pools of opportunity, hence in the way it prioritizes and allocates its resources. The improvement of the R&D process is pivotal and is resting both on expanding capabilities for external collaborations and streamlining internal processes to attract and retain key talents. Further, the industry should rebuild the social contract through a collaborative approach to value, building on engagement with the relevant stakeholders all along the life cycle of the novel drug.

The investors are expected to continue to fund the early stage of drug discovery, often through start-ups and biotech companies and to be prepared to fund innovative risk-sharing models. The dialogue with the industry should encompass both the full breadth of the R&D portfolio and the choice of R&D model, to have a comprehensive view of the drivers of the return on R&D investment.

The latest report published in 2012 extends this analysis by broadening the dialogue with other key stakeholders including payers, regulators, and societal and government experts. The report analyzes new market access trends (including the new regulations driving pricing and reimbursement) and the connection with innovation and productivity, and is essentially reinforcing the notions of integration that were present in the previous reports (Pharma Futures 5 2012).

The report underscores the importance of the health reforms, especially the indicators of accountability and health outcomes as essential tools to improve productivity and the greater importance placed on the patient as the pivot to determine clinical effectiveness and value for the health-care system. These reforms have an impact on pharmaceutical R&D, regulatory, market access and commercialization as they promote an approach based on more real-life evidence, generated at the patient level of by the patient herself. They create the conditions for "adaptive licensing," hence a step-wise and dynamic process towards the building of evidence of value rather than a binary one, and for in-depth collaborations along the key decision points, from the nature of the supportive data, to the access to innovation and to the pricing or reimbursement of the said innovation. The support of the investors is expected when they understand that such approaches open the possibility for earlier cash flows, and reduced regulatory and reimbursement risks.

Such a systemic approach will require transparent communication and frequent collaboration as a prerequisite for mutual understanding and trust building and as a way to overcome internal resistance and behaviors built over the past decades.

The same concepts have been the focus of a landmark article published in the Harvard Business Review by Michael Porter and Mark Kramer (Porter and Kramer 2011).

The authors are pointing that the business world has been criticized as a major cause of social, environmental, and economic problems and that companies are accused of prospering at the expense of their communities, resulting in an erosion of trust and in the enactment of policies that undermine competitiveness and sap economic growth. Further, they point that the companies are a major part of the problem, especially as they are focusing on the near-term financial returns rather than on the longer-term picture. The solution could rest in developing "shared value," i.e., financial value for the corporations and simultaneously value for society. Redefining R&D and the approach to the markets, availability and affordability, and supportive collaborations are at the core of the recommendation. The authors are mentioning a number of initiatives and argue that "shared value could reshape capitalism and its relationship to society." That would lead to the identification of new pools of potential for the companies and to the creation of a new set of competitive advantages, provided the companies evolve their set of skills and the authorities take a more educational and collaborative approach to regulation.

The perspective on shared value is evidently bordering that of protection of assets through intellectual property rights. As these have been blamed for being the obstacle to greater access to care and shared value, especially in emerging economies, authors such as Charles Leadbetter (Leadbetter 2009) have been suggesting solutions to address this specific question. The book tenet is about the value of "we" versus "I," or in other words, the new culture of sharing information and the value it creates in society at large and more specifically for some types of businesses. Regarding the health-care industry, it provides suggestions related to sharing knowledge for medical development and mentions Victoria Hale, who created the Institute for One World Health. The concept is to get pharmaceutical and biotech companies to donate patents and discoveries that they do not intend to develop and commercialize, but which may be of medical interest for the developing world. The institute subsequently assembles the right capabilities (scientific, financial) to develop these discoveries and bring them to where they bring value. This type of initiative raises the key question of ownership (not only objects but also knowledge, know-how, etc.) and drives home the point that "collaborative innovation invariably requires a form of shared ownership."

Sustainable Concerns All Along the Life Cycle of the Health-care Industry

As previously described, the health-care industry is one of long innovation cycles, resulting in a life cycle of 10–12 years or over, along which the question of green and sustainable pharmacy is always present (Kümmerer and Hempel 2010): Within recent years, pharmaceutical compounds have come under increasing scrutiny as contaminants of the environment and the issue of sustainable chemistry has gained momentum. The combination of the two is referred to as sustainable pharmacy, which is addressing environmental, economic, and social aspects of pharmacy.

The environmental dimension spans the entire life cycle of any pharmaceutical entity. It evidently included all the manufacturing questions, including the resources and energy consumption, and also the issues related to waste during the synthesis and production of an active pharmaceutical ingredient. Furthermore, it also considers the compounds themselves and aims to improve the biodegradability of the compounds after their human or veterinary use into the environment so as to reduce the risk caused by persisting chemicals. The approach also focuses on all the players along the prescription and dispensing value chain, as physicians, pharmacists, and patients respectively prescribe, dispense, and consume medications in a way that has a bearing on their presence in the environment. The question is to review their behaviors and assess how these could be more virtuous and contribute to more efficient use of pharmaceuticals with less environmental burden and less risk for drinking water. The book "Sustainable Pharmacy" addresses all these issues and is a pivotal piece dealing with this important topic.

The economic and social dimensions are multifaceted insofar as they tend to be polymorphic, depending on the geography and on the nature of the health-care ecosystem. Irrespective of the economic status of the said ecosystem, several authors have already elaborated on the importance of sustainable development and international collaboration in shaping the future of health-care systems, and how adequately funded R&D will assist in tackling current and predicted challenges (World Health Summit 2012). Across the entire wealth spectrum, the common challenge is the efficient management of scarce resources, ensuring that the gains of medical progress benefit as many people as possible. Related important topics span "priorities for research, public and private sector partnerships, intellectual property rights, regulatory procedures for health products, conventions on biomedical research and development and the place of information technology in health care systems".

> One of the biggest challenges for health in the globalized world is the privatization of the health sector and the lack of access of the poor to quality health services. Sima Samar (Chairperson of the Afghan Independent Human Rights Commission)

The challenge of managing diseases in modern environments is evidently an evolving one, as life expectancy has continued to rise steadily. Longer life expectancy, as well as unhealthy diets and sedentary lifestyles, has resulted in the global predominance of noncommunicable diseases as both the leading cause of death and of disease burden. Serious socioeconomic consequences can now be seen in both developed and developing countries, the latter often facing difficult arbitrage for resource allocation as they often have to deal simultaneously with the remnants of infectious diseases that are the hallmarks of poorer economies. These new issues are present on the global health agenda, alongside neglected diseases as well as future pandemics, for more balanced governance towards a healthier planet.

Public health interventions need to be designed and implemented taking into consideration research and innovation, as well as delivery of and access to that innovation, and capacity building and collaboration. As research is increasingly complex and technical, influencing policy and health-care practice requires transformation into usable information and a focus on a needs-driven research approach rather than market-driven approach, as it has been highlighted in the Global Strategies and Plan of Action of Public Health, Innovation, and Intellectual Property. Yet and as illustrated by the following quote, these matters go beyond the current scope of intervention of the health-care industry and require a broader collaborative approach, across different industry segments and coordinated by the public authority.

> To develop the political will for a health policy based on scientific data; in the tobacco epidemic, what works and what doesn't work has been proven for decades—it only requires political will to act in the short, medium and long-term interests of the health of the people. Judith Mackay (Senior Advisor, World Lung Foundation)

This is all the more important as the correlation between health and wealth has long been established, not only for emerging economies. Authors have described the impact of global financial crisis on health systems as "catastrophic" and are calling for more cooperation between the public and private sectors, and for sustaining investment and financing in health and social structures to maintain stability and security as well as to improve performance. This is the dimension where behavioral economics and the insights derived from data on consumer and lifestyle behaviors are expected to influence research and policy direction. As a consequence, drug development and usage will also be influenced, and this is already visible through new drug registration approaches (see interview of Richard Barker and the topic of adaptive licensing).

In the lower-income markets, access to health care—where available—has always been an issue, whether related to affordability, or awareness, and lower education level. As Severin Schwan (CEO, Roche Group) said "Lowering or removing these barriers is a shared responsibility, one that must be pursued more creatively and intensively in collaboration with health-care stakeholders worldwide—including governments, health-care providers and industry." Nowadays, all pharmaceutical companies have developed a strategic agenda for such sustainability initiatives in which educating health-care providers is a pivotal component. In addition, global health education will be essential to drive the required behavioral and consumption changes to maintain efficient health systems. As described in the previously referenced publication, "changing patterns of health threats in the twenty-first century such as those due to population movements and financial flows require a transformative educational approach of health professionals that are better attuned to the pressing needs for both global awareness and local sensitivity." The health-care industry is ideally positioned to contribute to the success of such educational initiatives as it hosts skilled resources and the knowledge of the patients and the diseases.

> It is against such a backdrop that one of the biggest challenges we face is getting people to work together—across agencies, governments, disciplines, and other boundaries, as well as changing human behavior. Although individual countries may be able to successfully develop strategies to counter some of the above, many global health issues defy borders and would require a collective strategy if we are to be successful. John Wong (Vice Provost (Academic Medicine) of the National University of Singapore)

As mentioned earlier, future success hinges upon collecting and analyzing massive amount of data and deriving actionable insights across the entire spectrum of health-care research, policy, and practice. The successful management of this information value chain will impact all the areas of health governance, research and innovation, politics and economics, as well as the education of health-care professionals and of the general population.

Factoring in all the above dimensions is an integral part of the strategic thinking in the health-care industry, all along its own value chain.

R&D

Much has been written about the evolution of the R&D output over the past decades and the conundrum that it represents for the health-care companies, the regulatory authorities, the payers, and the investors alike. The chapters presented in this book provide both an authoritative review and a fresh perspective of the critical topics of R&D productivity and patient centricity.

The chapter by *A. Schuhmacher* provides a comprehensive overview of the drivers of R&D sustainability. It analyzes where the pharmaceutical industry stands today in terms of innovation process and describes the drivers of the erosion of R&D productivity. Exploring the consequences of reduced R&D efficiency is then leading the author to recommend growth options to maintain sustainability for the health-care industry in the future, focusing on R&D-driven innovation.

In the chapter consolidating her executive insights, *K. Fischer* is demonstrating that capturing the patient voice early on in the development cycle is crucial to drug effectiveness. Based upon "real life" tradeoffs that patients are making around their treatments, she presents ways of using this data to improve the process of drug development, hence challenging the existing codes of practice, regulatory guidelines—originally designed to protect the patient—to ensure that their very importance voice is heard.

Another fascinating account of what can be accomplished when the industry gears itself properly to work more closely with the patients or the NGO representing them is given in the chapter presenting the interview of *S. Vink*. The insights are particularly relevant to the management of clinical development but also to global corporate governance as they address the need to evolve management practices, towards a long-term strategic orientation, openness to real-world data and responsiveness to societal pressure. The industry will need to flex its procedures, to open up to broader collaborative approaches and to foster a company-wide orientation towards innovation.

In the realm of R&D, other publications have addressed the broad topic of drug design and of "green chemistry," especially as compounds used in human medicine can cause adverse environmental effects. It is therefore argued that drug design should include consideration for environmental risk. In Sweden, systems for classification of drug environmental risk and hazard have been used for several years.

Although environmental data on human drugs are often missing, or reveal unfavorable environmental properties, it is argued that the pharmaceutical companies should highlight environmental precaution when designing new drugs (Wennmalm et al. 2008). Since chemical products are the main emissions of the pharmaceutical industry, it is difficult to hold them back efficiently. Very often they are not fully degraded to innocuous byproducts and unknown transformation products are formed in the environment. Publications have referenced case studies from industry, such as Taxol, Pregabalin, and Crestor, illustrating how a multidisciplinary approach to green chemistry has yielded efficient and environmentally friendly processes (Dunn et al. 2010). Therefore, according to the principles of green chemistry, the functionality of a chemical should not only include the properties of a chemical necessary for its application but also easy and fast degradability after its use. Authors advise taking into account the full life cycle of chemicals to lead to a different understanding of the functionality necessary for a chemical, factoring in its environmental properties. Several examples underline the feasibility and the economic potential of this approach, called benign by design (Kümmerer 2007).

This concept requires information on a compound's biodegradability to be available at an early stage, even before synthesis. Computer models for predicting biodegradation, therefore, are increasingly important, and various approaches to predict aquatic aerobic biodegradation have been critically reviewed from a user's point of view. The scientific debate addresses the fundamental problems in modeling biodegradation, as well as more general issues in modeling of compound properties by quantitative structure–property/activity relationships (Rücker et al. 2012).

The topic of preservation of natural resources has also been addressed in connection with R&D, exploring the connections between biodiversity, biotechnology, and sustainable development by examining the drug discovery process and agricultural improvements for better nutrition. Examples of ventures include the famous agreement between Merck & Co. and Costa Rica's National Institute for Biodiversity (INBio) and suggest policy options for potential host countries. The issues of costs, scientific and resource requirements, and economic prospects of different drug development models are also explored, as well as the combination of biodiversity and biotechnology to establish a sustainable agriculture. The authors have also delineated the legal ramifications of intellectual property rights, fair compensation for indigenous knowledge, and different contractual arrangements and more broadly how to assess biodiversity's economic value which could become the "green gold" and new competitive advantage for some countries (Pan American Sanitary Bureau 1996).

Manufacturing and Supply Chain

The sustainability of the production of human pharmaceuticals is multifaceted and includes topics that are common across to other industry sectors such as manufacturing constraints, concerns regarding the environment, knowing and managing the risks, wastewater treatment and energy consumption, as well as matters fully specific to the health-care industry such as technology transfer and north–south policies.

There is an abundant literature related to managing the risks for the environment and it is placed high on the agenda of the authorities. For instance, the US Environmental Protection Agency has released a list of 134 chemicals to be screened for their potential to disrupt the endocrine system of humans and animals, hence potentially affecting growth, metabolism, and reproduction.[22] Intensive research on pharmaceuticals in the environment started several years ago and a vast amount of literature has been published. The input and presence of active pharmaceutical ingredients (APIs) and their evolution in the environment remain of high interest. With the advent of proper measurement tools, it has been found that environmental concentrations can cause effects in wildlife and the question of mixture toxicity has gained more attention. Since work has been done in the field of risk assessment and risk management, the focus has been on raising discussions to influence policies in order to better manage risks (Kümmerer 2009b).

Energy saving has also received much attention, in the context of large manufacturing restructuring plans. One such example is Pfizer Germany GmbH's SPRING & E-MAP (Strategic Plant Restructuring & Energy Master Plan) project in Freiburg, Germany. The facility has won the Sustainability award in the 2011 Facility of the Year Award competition sponsored by ISPE, INTERPHEX, and Pharmaceutical Processing magazine.[23]

The pharmaceutical industry supply chain is subject to the Implementation Guidance Document (the Guidance Document) which conveys the spirit and intent of the Pharmaceutical Industry Principles for Responsible Supply Chain Management (the Principles) by providing a framework for improvement and examples of business practices and performance related to the principles.[24] This comprehensive document includes a description of the management systems required, such as the legal requirements, the tools of risk management, the required documentation, training and competencies and the need for continual improvement. The ethics section describes the principles of business integrity and fair competition, the identification of concerns, the principles of animal welfare and the privacy rules. The labor section contains a number of elements that are not specific to the health-care industry such as child labor, nondiscrimination and fair treatment, and the health and safety section expands on general topics such as worker protection, as well as on some that have aspects fully specific to the nature of the pharmaceutical products such as process safety, emergency preparedness, and hazard information. Evidently, the document elaborates on environmental protection, especially on the management of waste, emissions, and spills.

Technology transfer plays a pivotal role in the sustainable development activities related to the manufacturing of pharmaceutical products. As a matter of fact, trans-

---

[22] EPA to evaluate 134 chemicals for endocrine disruption http://www.environmentalleader.com/2010/11/17/epa-to-evaluate-134-chemicals-for-endocrine-disruption/. Accessed 10 Nov 2013.

[23] Pfizer discusses its strategic plant restructuring & energy master plan. http://www.environmentalleader.com/2011/04/11/pfizer-discusses-its-strategic-plant-restructuring-energy-master-plan/. Accessed 10 Nov 2013.

[24] Implementing the pharmaceutical industry principles for responsible supply chain management, pharmaceutical supply chain initiative. http://www.pharmaceuticalsupplychain.org/downloads/psci_guidance.pdf.

ferring technology contributes to improving the health of recipient countries' populations by facilitating access to innovative medicines and vaccines and strengthening local care procurement capacity. Such initiatives are usually part of broader programs including education of patients and populations at risk, and by conducting R&D on diseases specific to the developing world (IFPMA 2011).

Technology transfer is an accelerator of economic development as it allows emerging countries to access know-how and equipment relevant to the production of advanced health-care solutions. Once production is localized in an emerging market, it improves availability, access, and reliability of supply and creates high-tech job opportunities, therefore it contributes to improving the health and social status of the recipient country.

Many pharmaceutical companies have engaged successfully in technology transfer initiatives, including an educational component targeting the local health-care community and sometimes support to bettering the care procurement infrastructure. The importance of transferring technologies for medical products is recognized in the Global Strategy and Plan of Action on Public Health, Innovation, and Intellectual Property Rights of the World Health Organization (WHO). The technology transfers require a suitable local industrial partner to host the transferred technology. In addition, key success factors that have been described include a viable and accessible local market; political stability, good economic governance; clear development priorities; effective regulation; availability of skilled workers; adequate capital markets; strong intellectual property rights (IPR) and effective enforcement; and quality and duration of the relationship between industry and government.

Local governments create favorable conditions to attract technologies in demand by local manufacturers and evolve processes so as to foster mutual recognition of regulatory decisions. Technology transfer initiatives are also facilitated when they provide policy support for the development of a local private sector. Authorities in high income countries engage in educational initiatives to increase the technical expertise of the emerging countries regulators with the new technologies, while donors from developed economies provide funding for health care in the developing economies as a platform for development.

The IFPMA member companies are committed to transferring new technology and the relevant know-how and to delivering corporate social responsibility programs that offer products and specialized knowledge and skills contributing to economic development and public health of the recipient's country.

Licensing and Market Access

The current model for developing new drugs is becoming unaffordable, since costs to research and develop new drugs are steadily increasing, resulting in higher prices and mounting concerns among payers about affordability and cost-effectiveness and threatening access to novel therapies (Barker 2010). Obtaining a market authorization and a price or reimbursement has always been regarded by the industry as a complex process fraught with risks, prior to commercialization. The probabilities of success being variable, pharmaceutical manufacturers are usually planning for lengthy and complex negotiations. As the access to innovative medicine can be substantially delayed, the authorities have been taking this question very seriously.

Published literature has described the traditional drug licensing approaches as being based on binary decisions, insofar as at the moment of licensing, an experimental therapy is presumptively transformed into a vetted, safe, and efficacious therapy. Recently designed adaptive licensing (AL) approaches are based on stepwise acquisition of evidence, with iterative phases of data gathering and regulatory evaluation. The purpose of the approach is to align the content of the market authorization more closely with patient needs, and to enhance access to new technologies and to the evidence required to support medical decisions. Whether adaptive licensing is an evolutionary step or a transformative framework, it will inevitably require legislative action to create the conditions for routine implementation (Eichler 2012).

It is critical to understand the far-reaching implications of such adaptive approaches, since their implementation will impact the entire life cycle (research and clinical development, licensing, and market access) and require a wider breadth of collaboration by involvement of all stakeholders including the industry, regulator, payers/providers, and the research community. When successfully implemented, these approaches yield a specific clinical development plan that provides a staged access to evidence on risk versus benefit, subsequently enabling a faster review and expedited authorization in a well-defined group of patients. Along the clinical development and the commercial life of the product, the monitoring of "real-life" effectiveness and safety provides further evidence, driving the license adaptation.[25] The debate on adaptive licensing is still current, but the consensus among stakeholders is that the concept emerged from the necessity to react to the evolving pharmaceutical and economic context and that it should be pivotal to the future of pharmaceutical development and licensing by offering options for more flexible, adaptive, and collaborative design of the development and approval process (Barker and Garner 2012).

The concept of adaptive licensing and the consequences for the industry in terms of implementing a genuinely patient-centric strategy are described in the chapter by *R. Barker*. One of the most striking insight is that the industry appears as more conservative than the regulatory authorities when it comes to exploring new licensing routes and that a change of mindset is needed as much as an evolution of the industry's organization.

The required evolution encompasses a greater integration of industry functions along the product life cycle, the focus on better-profiled patient populations, taking into consideration both the clinical and behavioral dimensions, and a commitment to deliver the expected outcomes.

The chapter authored by *S. Chundru* provides the perspective on such regulatory matters from the perspective of the manufacturers originated in emerging countries.

---

[25] Strategy for UK life sciences-one year on, Project Director: Dr Sarah Garner. http://casmi.org.uk/adaptive-licensing/ (2012). Accessed 23 Mar 2014.

The author calls for a greater integration of the regulatory requirements in the clinical approach to licensing as well as for an improved harmonization of regulations across Western regulatory bodies which play a decisive role in sizing the business opportunity for emerging market industry players.

Consumption of Health-Care Products: Use, Access, and Disease Management

Towards the end of the life cycle of a pharmaceutical product, during the commercialization phase, the way the drugs are used is potentially raising sustainability questions ranging from disease management to elimination in the environment, the latter sharing roots with issues raised during research, development, and commercialization: since pharmaceuticals can be environmental pollutants, they require responsible use as much as novel testing and manufacturing approaches (Juniper 2013).

Regarding disease management, the chapter by F. Barei raises the key question of the cost at which the desired outcomes can be obtained. She underscores the pivotal role of education in evolving the health-care systems and of the broad span of supportive innovation, from the products' pharmaceutical presentation to the exploration of new business models, aiming at improving the patient's experience through convenience and adherence. She also underscores the foundational role of health-care IT and patient data.

The executive insights provided by V. Simons are especially relevant to understanding the patient perspective on sustainable health care in a patient-centric society. He addresses the topic from the dual perspective of the patient and the founder of a patient association aiming at educating consumers most at-risk from a diagnosis of prostate cancer, informing the community on other diseases and conditions of negative impact, motivating consumers to make informed choices as to health-care and lifestyle management, laying the foundation for ongoing health-care information dissemination and interaction between the community and medical centers and creating an interactive network to maximize broad scale, mass communications of actionable health messages.[26] He provides examples of partnerships which contributed to improve drug development and industry processes, and also access to care, by focusing on the patient perspective.

Understanding the question of elimination of medications: The primary route by which active ingredients from human pharmaceutical products enter the environment is excretion in urine and feces. Besides, the disposal of unwanted, leftover medications by flushing into sewers also adds to environmental pollution but is considered to be of a lesser importance. Authors have argued that understanding these secondary routes is important from the perspective of preventing such pollution, because actions can be designed more easily for reducing the environmental

---

[26] http://www.theprostatenet.org/aboutUs.html. Accessed 15 Aug 2014.

impact of active ingredients compared with the route of direct excretion (via urine and feces). The expected benefits should include the reduction of the incidence of unintentional poisonings of humans and animals, and the improvement of the quality and cost-effectiveness of health care, since the unintentional exposure to active ingredients for humans via these routes is possibly more important than exposure to residues recycled from the environment in drinking water or foods (Daughton 2009, Daughton et al. 2009).

A review of the challenges posed by antibiotics in the aquatic environment is an illustration of the magnitude of the issue and of its far-reaching implications. Antibiotics have been used extensively for decades, yet the existence of these substances in the environment has gained attention only recently, with a detailed assessment of the environmental risks they may pose. Within the last decade, an increasing number of studies covering antibiotic input, occurrence, fate and effects have been published, but there is still a lack of understanding and knowledge about antibiotics in the aquatic environment despite the numerous studies performed. Important questions are still open, especially the risks associated with antibiotics presence in the environment, such as bacterial resistance (Kümmerer 2009a).

Even the prescribed use of pharmaceuticals can result in unintended, unwelcomed, and potentially adverse consequences for the environment and for those not initially targeted for the treatment. Depending on the nature of the active ingredient, medication usage frequently results in the collateral introduction to the environment of the said active ingredients or bioactive metabolites, and reversible conjugates. As mentioned earlier, imprudent prescribing and noncompliant patient behavior drive the accumulation of unused medications, which can pose major public health risks from diversion as well as risks for the environment when disposed inappropriately. The prescribers very seldom incorporate consideration of the potential for adverse environmental impacts into daily prescribing practice. Prescription guidelines could encourage the selection of medications possessing environment-friendly excretion profiles and the prescription of the lowest effective dose suiting the patient needs, reducing the incidence of adverse drug events and lowering health-care costs. It is argued that the prescriber needs to be cognizant that the "patient" encompasses the environment and other "bystanders," and that prescribed treatments can have unanticipated, collateral impacts that reach far beyond the health-care setting (Daughton et al. 2013).

Examples of initiatives of health-care companies taking the broader perspective of the patient in the environment have been published, such as GlaxoSmithKline's recycling of asthma inhalers. The initiative targets the collection of 100,000 used respiratory inhalers through a program to make new household products, such as plastic hangers and plastic flowerpots.[27] Also, Johnson & Johnson had 30 products in its Earthwards portfolio of environmentally conscious health-care and pharmaceutical products in 2010; it added 19 products to the range in 2011. In 2012, J&J

---

[27] GlaxoSmithKline aims to collect 100,000 inhalers. http://www.environmentalleader.com/2012/10/25/glaxosmithkline-aims-to-collect-100000-inhalers/. Accessed 10 Nov 2013.

was halfway towards its goal of having 60 products in the line by 2015.[28] Companies are also benefiting by leveraging "green" products as part of their payer value proposition. For instance, 35 % of hospitals surveyed switched suppliers to gain access to sustainable health-care products; according to a 2012 report by Johnson & Johnson that finds hospitals to be placing greater emphasis on "green" products used in patient care and throughout the facilities such as cleaning products (Johnson & Johnson 2012).

Beyond the questions raised by the individual use of medications, the pharmaceutical companies have created competitive advantages by addressing unmet medical needs and access to care in low- and middle-income countries (Peterson et al.). The execution of this strategy often rests on partnerships, such as the 213 programs (recorded from 2003 to 2010), in the fields of HIV/AIDS, tuberculosis, malaria, and tropical diseases as well as other health needs, including preventable diseases, child and maternal health, chronic diseases, and additional health initiatives. The research-based health-care industry contributes to strengthening overall health care by implementing access and capacity-building programs in developing countries (IFPMA 2010).

The insights presented in the chapter transcribing the interview of *E. Pisani,* Director General of IFMPA, help put things in the global perspective by underscoring the pivotal value of global partnerships in the virtuous circle of reinforcing both health and wealth.

Global health partnerships play a pivotal role in meeting many of the most critical health needs of low- and middle-income countries[29]. Global health partnerships (GHPs) have evolved to become an effective vehicle for collaboration to address global health challenges. The BSR report summarizes the contribution of GHPs to meeting global health needs with a focus on low- and middle-income countries and provides perspectives on how to increase the impact and scale of GHPs going forward, based on interviews with leaders from the private sector and stakeholder groups, an assessment of more than 220 partnerships, a survey of pharmaceutical industry executives, and a multi-stakeholder roundtable convened in Geneva in December 2011.

Examples include the HIV/AIDS partnerships aiming at creating pediatric treatment centers, training health-care professionals, and working with community implementation partners to reduce stigma, promote prevention, increase rates of diagnosis, and to assist patients to comply with treatment regimens. The report also highlights Malaria-focused partnerships—and others focused on tropical diseases—which are facilitating technology transfer agreements (see the above section, related to R&D) for research on new compounds, training community health workers, providing education and outreach on prevention, enabling donations and differential pricing arrangements for no- and low-cost medication, and providing professional

---

[28] Johnson & Johnson expands green product range. http://www.environmentalleader.com/2012/03/02/johnson-johnson-expands-green-product-range/. Accessed 10 Nov 2013.

[29] Working toward transformational health partnerships in low and middle income countries, BSR; 2012.

education and best practice sharing for health-care professionals and policy makers. Among the partnerships surveyed, those focused on HIV/AIDS represent 20%, malaria accounts for 14%, and neglected tropical diseases for 16%, hence a total of 50% of the total partnerships surveyed. Only 14% of all GHPs focus on noncommunicable diseases. Concurring with this analysis, input from stakeholders and companies alike confirm that there is an increasing need for GHPs to focus on the unique challenges presented by NCDs in developing countries.

The chapter presenting the highlights of the interview with *F. Bompart* provides a detailed account of the tools that are part of the corporate responsibility programs, such as tiered pricing in the specific context of a malaria-focused initiative, and explores the recent evolutions since the early programs which were derived from HIV politics and market dynamics.

As the needs of the emerging countries are evolving, partnerships also target noncommunicable diseases and contribute to primary health systems that provide the foundation for diagnosis and continuous care across a range of chronic diseases. For instance, regarding diabetes and cardiovascular diseases, insufficient capacity of primary health systems poses a critical challenge to diagnosis and patient management. In the absence of a robust primary care system, populations are often underdiagnosed and untreated until the disease state generates complications, and requires more challenging (and expensive) treatment regimens, and raises the threat of reduced life expectancy.

The disease areas addressed by partnerships are broad, and the challenge for companies investing in such partnerships is to evolve so as to address the epidemiology shift towards chronic, noncommunicable diseases. More specifically, much remains to be accomplished in terms of addressing diagnosis, treatment, and managed care for such diseases in low-resource environments. At the same time, companies must maintain the legacy partnerships (e.g., HIV and malaria) where continued investment is critical to ensuring long-term disease control. This prioritization is required as well as an overall increase in allocated resources.

The challenges for companies (and the expectation of shareholders) are to identify indicators allowing measuring the impact of such initiatives. Tracking medical outcomes is often difficult due to lack of data and analytic resources both at the local level as well as within NGOs involved, let alone identifying overarching impact such as workforce productivity. Some of these challenges were attributed simply to a lack of resources allocated to unlocking this difficult puzzle. More development work is needed to expand the set of indicators and generate insights on the total impact achieved by partnerships. In the long-run, impact measurement should move from measurement of activity indicators (e.g., number of physicians trained) to highlighting performance indicators on wellness and life expectancy.

High-impact partnerships (or "transformational partnerships") are described as cutting across therapeutic areas, building primary-care systems, and developing local capacity for prevention, diagnosis, and treatment across a full range of diseases.

For the time being, most partnerships involve a single company working with a variety of partners, including NGOs, governments, and academics. These partnerships have made significant contributions to global health in terms of the range of perspectives and diverse approaches to global health challenges. The specific challenges posed by non-communicable diseases raise the prospect of a higher proportion of partnerships involving several research-based companies to capitalize on a broader range of expertise and assets, most importantly funding, products, R&D capabilities, and skills and time of dedicated staff. Funding from companies is not fully addressing health-care needs in low- and middle-income countries, and the concern is that the public funding is declining throughout the world, further fueling the debate about the role of the private sector in driving global health outcomes. When corporate strategic orientations are aligned with the need to increase investment in partnerships, there are opportunities to develop innovative approaches for internal resources in ways that build local capabilities and simultaneously serve the company's increasing need to increase its intimacy with local markets.

Global Corporate Governance

As argued by Ian Davies, "By building social issues into strategy, big business can recast the debate about its role" (Davis 2005). The UN Global Compact requested to evaluate the potential for corporate responsibility initiatives to stimulate a transition to more sustainable forms of development by linking to wider policy frameworks.[30] Sustainable development in both the developed and developing world revolves around the common fundamental themes of advancing economic and social prosperity while protecting and restoring natural systems. For the time being, the majority of initiatives have focused on transferring knowledge from the developed to the developing world, yet there is an increasing body of evidence suggesting that indigenous knowledge from developing countries can contribute to the global dialogue, especially in the critical fields of water and energy. Case studies demonstrate that, with relevant analysis and quantification, insights can be adapted for transfer throughout the developed and developing world in advancing sustainability, especially the integration of natural processes and material flows into the anthropogenic system. As the global trend of urbanization accelerates, innovations applied to water and energy are expected to fundamentally shift the type and efficiency of energy and materials utilized to advance prosperity while protecting and restoring natural systems (Mihelcic et al.). Interestingly, despite a growing number of bold and visionary companies making considerable achievements, the overall corporate impact on critical sustainability issues—such as sanitation, health care and climate change—has been limited. The key will be to scale up corporate responsibility initiatives to make a greater contribution to addressing global challenges.[31]

---

[30] Gearing Up: From corporate responsibility to good governance and scalable solutions, http://www.sustainability.com/library/gearing-up?path=gearing-up#.UTYBhYlespo. Accessed 23 Mar 2014.

[31] Issue Brief: Progressive alliances, scaling up corporate responsibility to address global challenges http://www.sustainability.com/library/issue-brief-progressive-alliances#.UTYAn4lespo. Accessed 23 Mar 2014.

The chapter authored by *G.C. Chen* investigates how the biomedical enterprises transform their strategic goals to fulfill the mission of sustainability and addresses the need of a customized framework for biomedical enterprises. Based on an analysis of the current practices on the issues of sustainable development relative to the size of the corporations, the author provides recommendations about how to improve strategies for sustainable development.

The perspective from a health-care company based in a mature market is captured in the chapter by *V. Logerais*. She places the topic in perspective with both the moral obligations and the regulatory constraints weighing on the business and highlights the steps taken by the company to define the scope of its accountabilities and the operational implications. These include the environmental impact of the products and their manufacturing, purchasing principles compliant with fair trade guidelines, and a societal policy targeting the company employees. The chapter provides a clear example of an integrated company policy towards sustainable development that is relevant to the strategic situation and orientations.

A review of existing corporate sustainability policies demonstrates that companies are investing strongly as they generate evidence that such investment creates significant shareholder value, in a measurable way. The following examples focus on health-care companies.

In 2011, GlaxoSmithKline has set a target to achieve carbon neutrality across its value chain by 2050, as part of a new environmental strategy launched in the company's 2010 corporate responsibility report. The carbon neutrality target means that within 40 years, there will be no net greenhouse gas emissions from GSK's raw material sourcing, manufacturing, distribution, product use and disposal, the company said. It has set interim targets to reduce its carbon footprint by 10% by 2015 and 25% by 2020.[32]

In 2010, Pfizer reported average savings of US$ 1.4 million annually between 2004 and 2009 by installing energy-efficient light fixtures, timers, and occupancy sensors at all of its Kalamazoo, Mich., facilities. Savings in 2009 alone tallied US$ 2.6 million.[33]

Also in 2010, the campus of Janssen, Division of Ortho-McNeil-Janssen Pharmaceuticals, Johnson & Johnson flipped the switch on the largest solar panel array in New Jersey—as well as the largest solar installation of any site among the Johnson & Johnson family of companies.[34]

Novo Nordisk commented in its 2010 integrated annual report on the outcomes of sustainable initiatives that it exceeded long-term targets for reducing CO2 emissions, water consumption, and total energy consumption, while increasing its work-

---

[32] GlaxoSmithKline sets carbon neutrality goal for 2050 http://www.environmentalleader.com/2011/03/30/glaxosmithkline-sets-carbon-neutrality-goal-for-2050/. Accessed 10 Nov 2013.

[33] Energy-efficient measures saves Pfizer $ 2.6M in 2009 http://www.environmentalleader.com/2010/10/07/energy-efficient-measures-saves-pfizer-2-6m-in-2009/. Accessed 10 Nov 2013.

[34] Johnson & Johnson COMPLETES LARGEST SOLAR PANEL ARRay in NJ http://www.environmentalleader.com/2010/09/22/johnson-johnson-completes-largest-solar-panel-array-in-nj/. Accessed 10 Nov 2014.

force by 8% and sales by 12% in 2009. The company reduced CO2 emissions by 32% and water consumption by 20% in 2009.[35]

In 2009, Eli Lilly, reportedly reached its energy goal 2 years early in the context of a long haul program, improving its energy intensity (energy used per dollar of sales) by more than 35% and cutting its absolute energy use by 5.8% from 2004 to 2008. Over the same period, the company also cut its absolute greenhouse gas (GHG) emissions by 4.4%.[36] In the same year, Pfizer announced in its corporate sustainability report that it achieved three of its four public environmental goals to reduce emissions, also from a long-term initiative. The company exceeded its goal to reduce greenhouse gas (GHG) emissions by 35% on a relative basis from 2000 to 2007, cutting emissions by 43% in 2007 and an additional 20% over 2007 to 2008.[37]

A broad and benchmarked perspective is provided by consultancies that are conducting research and comparing company programs. Bristol-Myers Squibb and Sanofi-Aventis earned the two highest scores in the pharmaceutical sector for sustainability reporting, according to a 2009 report from the Roberts Environmental Center.[38] In the 2012 issue of the report, the highest marks went to Merck, Amgen and Abbott[39], indicating that the momentum is gaining the entire industry but that much remains to be done, since some of the score vary very substantially between the top and the bottom performers.

Conclusion

Health-care markets are evolving under demographic and economic pressures. In mature markets, patients navigate highly complex provider and payer systems with limited control on health-care quality and outcomes, as reflected by the absence of correlation between spending and patient satisfaction and outcomes. In developing markets, patients have limited—yet growing at varying paces—awareness, access, and ability to pay for health care. The per capita health-care spending is directly correlated to GDP growth, driving significant expansion of health-care markets

---

[35] Novo Nordisk Cuts CO2 Emissions by 32%, Water Use by 20% http://www.environmental-leader.com/2010/02/08/novo-nordisk-cuts-co2-emissions-by-32-water-use-by-20/. Accessed 10 Nov 2013.

[36] Lilly meets energy-efficiency goals ahead of schedule. http://www.environmentalleader.com/2009/10/23/lilly-meets-energy-efficiency-goals-ahead-of-schedule/. Accessed 10 Nov 2013

[37] Pfizer exceeds emissions reduction goals, misses clean energy target. http://www.environmentalleader.com/2009/10/07/pfizer-exceeds-emissions-reduction-goals-misses-clean-energy-target/. Accessed 10 Nov 2013.

[38] Bristol-Myers Squibb, Sanofi-Aventis tops pharmaceutical sustainability report. http://www.environmentalleader.com/2009/12/15/bristol-myers-squibb-sanofi-aventis-tops-pharmaceutical-sustainability-report/?graph=full&id=1. Accessed 10 Nov 2013

[39] https://www.claremontmckenna.edu/roberts-environmental-center/wp-content/uploads/2014/02/Pharmaceuticals2012.pdf. Accessed 15 Aug 2014.

in emerging countries. The largest part of health-care spending (estimated around 50% by 2020) will be driven by providers and services, while the industry will represent about 10%. The pools of profitability will progressively shift from prescription drugs to other product segments and to health-care delivery, and these shifts will be different by region.

The health-care industry needs to identify which businesses are attractive for major or new investments, in a sustainable manner. Sustainable growth in the health-care industry has already been extensively investigated and analyzed and this has left a trail of publications which focus primarily on R&D productivity (especially the ability to maintain a stream of innovation to replace the top selling drugs which lose exclusivity, the management of the R&D risk through a balanced portfolio, etc.), on tight management of the cost structure (particularly for those health-care industries which are capital intensive, such as the biopharmaceutical industry), on commercial effectiveness (for instance looking at the evolution of the marketing mix and the shift in analytics and marketing tactics to maximize return on promotional investments), etc.

The topic of sustainable development raises additional questions, some being specific to the health-care industry such as:

- The preservation of rare natural resources (leading to questions such as equitable prospecting of specific plants, sourcing, and respect of biodiversity)
- The performance or eco-efficiency (which can be specific to the health-care industry for its most capital-intensive segments such as the biological produced at large scale, such as vaccines)
- The environmental effects (especially pollution and hazards created by the handling of toxic or infectious materials, at industrial scale)
- The capacity issues in relation to the size of the medical need and/or of the market demand (leading to issues of access to care, especially in low- to middle-income markets)
- The recycling strategies for biopharmaceutical processes
- The private–public partnerships to address global health crises, and which have a bearing on the image management policies (especially in a global context of a growing challenge of the "for profit" model of the health-care industry)

The products researched, developed, manufactured, and commercialized by the health-care industry focus on improving the health of patients and populations, and regulations are applied to ensure the security of the end users and the sustainability of the health-care systems. The standardization of health-care products (both small molecules and biologicals) is de facto a driver of sustainable development, insofar as the search for reproducibility drives manufacturers away from the variability-induced extraction of compounds from natural resources towards, the use of materials of animal origin, etc., towards better defined chemical synthesis or bio-fermentation processes.

Overall, the entire value chain of the health-care industry should be engaged in the execution of the corporate strategic goals aiming at enhancing sustainability. Among the levers allowing to balance decisions related to risk management with long-term shareholder value and business sustainability, patient centricity along the entire life cycle, integration of data and company functions, and engagement and collaboration with a broader scope of stakeholders have been demonstrated to be the most effective.

We expect that the present book will encourage academics, industry specialists, and representatives of the civil society to devote further attention to research on performance indicators generating supportive evidence of long term, shared value of sustainable development endeavors.

References

Barker R. A flexible blueprint for the future of drug development. Lancet 2010;375:357–9.

Barker R, Garner S. Adaptive drug development and licensing. Regulatory Rapporteur 2012;9, 10:13–5.

Baumol WJ. The cost disease, why computers get cheaper and health care doesn't. Yale University Press, USA; 25 Sept 2012. ISBN: 9780300179286.

Daughton CG. Chemicals from the practice of healthcare: challenges and unknowns posed by residues in the environment. Environ Toxicol Chem. 2009;28(12):2490–4. http://onlinelibrary.wiley.com/doi/10.1897/09-138.1/abstract. Accessed 3 Aug 2014.

Daughton CG, et al. Environmental footprint of pharmaceuticals: the significance of factors beyond direct excretion to sewers. Environ Toxicol Chem. 2009;28(12):2495–521. http://onlinelibrary.wiley.com/doi/10.1897/08-382.1/full. Accessed 3 Aug 2014.

Daughton CG, et al. Lower-dose prescribing: minimizing "side effects" of pharmaceuticals on society and the environment. Sci Total Environ. 2013;443:324–37.

Davis I. The biggest contract. The Economist; 26 May 2005. http://www.economist.com/node/4008642. Accessed 10 Nov 2013.

Dunn PJ, Wells A, Williams MT, editors. Green chemistry in the pharmaceutical industry. 2010 (ISBN: 978-3-527-32418-7).

Eichler HG, et al. Adaptive licensing: taking the next step in the evolution of drug licensing. Clin Pharmacol Ther. 2012;91(3):426–37. doi:10.1038/clpt.2011.345 (Epub 15 Feb 2012).

Greider W. The soul of capitalism—opening path to a moral economy. Simon & Schuster, New York; 2003.

IFPMA. Developing world health partnerships directory. IFPMA Report; 2010.

IFPMA. Technology transfer: a collaborative approach to improve global health. IFPMA Report; 2011.

Johnson & Johnson. 54 % of hospitals say green attributes important in purchasing. Report. http://www.environmentalleader.com/2012/09/26/johnson-johnson-report-54-of-hospitals-say-green-attributes-important-in-purchasing/ (2012). Accessed 10 Nov 2013.

Juniper T. What has nature ever done for us? Profile Books, London, UK; 2013 (ISBN 978-1-84668-560-6).

Kümmerer K. Sustainable from the very beginning: rational design of molecules by life cycle engineering as an important approach for green pharmacy and green chemistry. Green Chem. 2007;9:899–907. doi:10.1039/B618298B (Received 14 Dec 2006, Accepted 15 Mar 2007, First published on the web 05 Apr 2007).

Kümmerer K. Antibiotics in the aquatic environment—a review—Part I. Chemosphere 2009a;75(4):417–34.

Kümmerer K. The presence of pharmaceuticals in the environment due to human use—present knowledge and future challenges. J Environ Manag. 2009b;90(8):2354–66.

Kümmerer K, Hempel M, editors. Green and sustainable pharmacy. 1st ed. Vol. XVII, Springer, New York, USA, p. 313 40 illus. 2010. ISBN 978-3-642-05199-9.

Leadbetter C. We think. Profile Books Ltd, London, UK (ISBN 978-1861978370); 2009.

Mihelcic JR, et al. Integrating developed and developing, world knowledge into global discussions and strategies for sustainability. 1. Science and technology. Environ Sci Technol. 2007; 41(10):3415–21.

Mistra Pharma: A healthy future—pharmaceuticals in a sustainable society. 2009 (Published in collaboration between Apoteket AB, MistraPharma and Stockholm, Sweden County Council).

Pan American Sanitary Bureau. Biodiversity, biotechnology, and sustainable development in health and agriculture. 1996 (ISBN: 92 75 11560 5)

Paul SM, et al. How to improve R&D productivity: the pharmaceutical industry's grand challenge. Nat Rev Drug Discov. 2010;9:203–14.

Peterson K, et al. Competing by saving lives: how pharmaceutical and medical device companies create shared value in global health, 2012—report http://www.fsg.org/tabid/191/ArticleId/557/Default.aspx?srpush=true. Accessed 23 Mar 2014.

Pharma Futures 1. The pharmaceutical sector: a long-term value outlook. 2004. http://pharmafutures.org/pharmafutures-1/a-long-term-value-outlook/. Accessed 23 Mar 2014.

Pharma Futures 2. Prescription for long-term value. 19 Jun 2007—Report http://www.sustainability.com/library/pharma-futures-2#.UTYG04lespo. Accessed 23 Mar 2014.

Pharma Futures 3. Emerging opportunities-23 Feb 2009—Report http://pharmafutures.org/pharmafutures-3/emerging-opportunities/. Accessed 23 Mar 2014.

Pharma Futures 4. Shared value, 2011—report. http://pharmafutures.org/pharmafutures-4/shared-value/. Accessed 23 Mar 2014.

Pharma Futures 5. Innovation and productivity in health systems, 2012—report. http://pharmafutures.org/pharmafutures-5/innovation-in-health-systems/. Accessed 23 Mar 2014.

Porter M, Kramer M. Creating shared value. Harv Bus Rev. 2011;89(1–2):62–77.

Rücker C, et al. Modeling and predicting aquatic aerobic biodegradation—a review from a user's perspective. Green Chem. 2012;14:875–87. doi:10.1039/C2GC16267A.

Wennmalm A, et al. Drug design should involve consideration of environmental risk and hazard. Lett Drug Des Discov. 2008;5(4):232–5.

Wennmalm A, et al. The vision—sustainable pharmaceutical management in a sustainable society. MistraPharma, April 2010, p. 130.

World Economic Forum. Sustainable health systems: visions, strategies, critical uncertainties and scenarios. January 2013.

World Health Summit 2012. Research for health and sustainable development, M8 Alliance; 2012.

Chapter 2
Can Innovation Still Be the Main Growth Driver of the Pharmaceutical Industry?

Alexander Schuhmacher

Innovation as a Driver of Growth for the Pharmaceutical Industry in the Past

In the period from the 1950s to 2013, the American Food and Drug Administration (FDA) approved 1346 new molecular entities (NMEs) or new biologics entities (NBEs). On average, the approval rate was 20 NMEs per year. In the past 40 years, the number of new drugs launched into the market increased slightly from 15 NMEs in the 1970s to 25–30 NMEs since the 1990s (Munos 2009). The highest number of new drugs approved by FDA was in 1996 and 1997 (see Fig. 2.1), which might be related to the enactment of the Prescription Drug User Fee Act (PDUFA) in 1993 (Kaitin and DiMasi 2011).

It has been reported that in 2009 approximately 4300 pharmaceutical companies performed research and development (R&D) worldwide (Munos 2009). Compared to this figure, it is interesting to note that from 1950 to 2009 only 261 pharmaceutical companies have been successful in launching at least one new drug into the market (Munos 2009). Out of this group, only 12 % of the companies were in the pharmaceutical market for all 60 years (Munos 2009). The other organizations either failed, merged with a competitor, or were acquired. About 600 NMEs were launched by the companies that disappeared due to merger and acquisition (M&A; Munos 2009). Twenty-one pharmaceutical companies have launched 50 % of all new drugs until today, whereby 360 NMEs have been produced by nine pharmaceutical companies that have existed since 1950 (Munos 2009). Out of this group, Merck & Co. (www.merck.com), Eli Lilly (www.lilly.com), and Roche (www.roche.com) have been the most successful companies worldwide so far (Munos 2009). The fact that some companies were able to survive over a period of six decades shows that the health-care sector has provided a basis for the sustainable growth of pharmaceutical

---

A. Schuhmacher (✉)
Reutlingen University, Reutlingen, Germany
e-mail: alexander.schuhmacher@reutlingen-university.de

© Springer International Publishing Switzerland 2015
P. A. Morgon (ed.), *Sustainable Development for the Healthcare Industry*,
Perspectives on Sustainable Growth, DOI 10.1007/978-3-319-12526-8_2

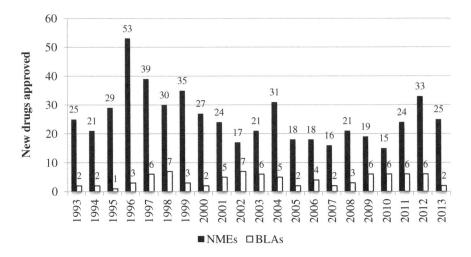

Fig. 2.1 New drugs approved by FDA between 1993 and 2013. (Data derived from Hughes 2009; Munos 2009; Mullard 2012b, 2014b; www.fda.org); NMEs new molecular entities, BLAs biologic license applications, FDA Food and Drug Administration

companies up to this point. But is there also ground for future growth and sustainability for pharmaceutical companies in the future?

The Pharmaceutical Industry Today

The R&D Investments of Top Pharmaceutical Companies

Today, the multinational pharmaceutical companies that perform R&D come from the traditional, main pharmaceutical markets, namely the USA, Europe, and Japan. Of the 15 companies listed in Table 2.1, seven companies are based in the USA, two in Japan, and six in Europe. None of these major players in the pharmaceutical industry come from emerging countries such as China, India, Russia, Brazil, or South Africa.

The pharmaceutical sector is still polypolic. The top 15 pharmaceutical companies have a combined market share of 51.8%. Today's leading pharmaceutical company worldwide is the Swiss Novartis with total group sales of US$ 50.8 billion in 2012. Its R&D investments have been enormous in recent years with the totals of US$ 8–9 billion annually (see Table 2.2).

On average, the top pharmaceutical companies have invested 15–20% of their total sales into R&D in the past years, which has translated into R&D costs of more than US$ 5 billion annually (see Table 2.3). The overall average R&D rate of

Table 2.1 Top pharmaceutical companies ranked in accordance with their total pharmaceutical sales in 2012. Not included are revenues generated by nonpharmaceutical activities

Rank	Company	Headquarter (city, country)	Total sales (USD billion, 2012)	Market share (%)
1	Novartis	Basel, CH	50.8	5.9
2	Pfizer	New York, USA	46.9	5.5
3	Merck & Co.	Whitehouse Station, USA	40.2	4.7
4	Sanofi	Paris, FR	37.7	4.4
5	Roche	Basel, CH	34.8	4.1
6	GlaxoSmithKline	Brentford, GB	32.7	3.8
7	AstraZeneca	London, GB	32.0	3.7
8	Johnson & Johnson	New Brunswick, USA	27.9	3.3
9	Abbott	North Chicago, USA	26.8	3.1
10	Teva	Petach Tikwa, IS	24.8	2.9
11	Eli Lilly	Indianapolis, USA	21.9	2.6
12	Amgen	Thousand Oaks, USA	17.2	2.0
13	Boehringer Ingelheim	Ingelheim, DE	17.1	2.0
14	Bayer	Leverkusen, DE	16.2	1.9
15	Takeda	Osaka, JP	15.9	1.9

USD US Dollars

the pharmaceutical and biotechnology industry has been described to be 14.4 % in 2012 (European Commission 2013). Companies such as Novartis, Pfizer, Roche, and Sanofi have even invested more than US$ 8 billion per year showing the importance of R&D as a major driver of growth in the industry.

According to the European Commission, 15 of the top 50 companies that invest most in R&D worldwide are pharmaceutical companies (European Commission 2013). Thus, the pharmaceutical branch is one of the top investors in R&D worldwide. Roche (6), Novartis (7), Merck & Co. (8), Johnson & Johnson (9), and Pfizer (10) are within the top ten of the world leading R&D investors (European Commission 2013).

In total, the pharmaceutical industry is the sector that invests most in R&D worldwide. The International Federation of Pharmaceutical Manufacturers and Associations (IFPMA) reported that in 2010 the pharmaceutical and biotechnology industries had R&D investments of more than US$ 85 billion (IFPMA 2012) with US$ 48.5 billion R&D investments reported by Pharmaceutical Research and Manufacturers of America (PhRMA) members (PhRMA 2013).

Resulting from increasing R&D expenditures during the years 2005–2012, the European Commission reported an investment in R&D of up to US$ 100 billion

Table 2.2 R&D investments and R&D rate of Novartis (2001–2013). R&D rate is the relative proportion of R&D costs to total sales per year

Year	Novartis		
	Total sales (USD million)	R&D costs (USD million)	R&D rate (%)
2001	32.038	4.189	13.1
2002	20.877	2.843	13.6
2003	24.864	3.765	15.1
2004	28.247	4.207	14.9
2005	29.400	4.800	16.3
2006	34.400	5.300	15.4
2007	38.100	6.400	16.8
2008	41.500	7.200	17.3
2009	44.300	7.300	16.5
2010	50.600	8.100	16.0
2011	58.600	9.200	15.7
2012	56.700	9.100	16.0
2013	57.900	9.600	16.6

USD US Dollars

worldwide for the pharmaceutical and biotechnology sectors in 2012 (European Commission 2013). In the same report, the analysis showed that most of the multinational pharmaceutical companies have invested significantly more in R&D during the period between 2005 and 2012 (see Table 2.4; European Commission 2013).

The huge amounts pharmaceutical companies are spending in new drug R&D and the enormous total R&D investments of the whole industry have put pressure on the return on R&D investment and brought the sustainability of pharmaceutical R&D in question if the output, namely the number of new drugs launched, is not comparably high.

The Output of Pharmaceutical R&D

In the past 12 years, Novartis (www.novartis.com), Pfizer (www.Pfizer.com), and GlaxoSmithKline (www.gsk.com) have been the most successful pharmaceutical companies, as they launched 16, 13, and 12 new drugs into the market, respectively. Figure 2.2 summarizes the number of NMEs from the most efficient pharmaceutical companies that have been approved by the FDA over the period of 2001–2012.

The statistics of new drugs launched into the market in Fig. 2.2 include the NMEs per company that have been generated from internal sources and also the ones that come from external sources, such as licensing of drug candidates and acquiring new drugs by M&A. The total externally sourced pipeline of multinational pharmaceutical companies has been analyzed to be sourced on average by 50% (29–80%)

Table 2.3 Key R&D figures of the top pharmaceutical companies

Year	AstraZeneca		GSK	Merck & Co.		Pfizer		Roche		Sanofi	
	R&D costs (USD million)	R&D rate (%)	R&D costs (BP million)	R&D costs (USD million)	R&D rate (%)	R&D costs (USD million)	R&D rate (%)	R&D costs (CHF million)	R&D rate (%)	R&D costs (EUR million)	R&D rate (%)
				R&D rate (%)							
2005	3379	14.1	3136	14.5 / 3848 / 17.5		7256	15.3	5705	16.1	4044	14.8
2006	3902	14.7	3457	14.9 / 4783 / 21.1		7599	15.7	7365	17.5	4430	15.6
2007	5089	21.6	3327	14.6 / 4883 / 20.2		8089	16.7	8385	18.2	4537	16.2
2008	5179	16.4	3681	15.1 / 4805 / 20.1		7945	16.5	8845	19.4	4150	16.8
2009	4409	13.4	4106	14.5 / 5845 / 21.3		7845	15.7	9874	20.1	4091	15.8
2010	5318	16.0	4457	15.7 / 11,111 / 24.2		9413	13.9	9050	19.1	3884	14.6
2011	5523	16.4	3687	13.5 / 8467 / 17.6		8681	14.2	8073	19.0	4101	14.7
2012	5243	18.7	3979	15.1 / 8168 / 17.3		7482	13.7	8475	18.6	4905	14.0
2013	4821	18.8	3923	14.8 / 7503 / 17.0		6678	12.9	8700	18.6	4770	14.5

GSK GlaxoSmithKline, *CHF* Swiss Franks, *BP* British Pounds, *USD* US Dollars, *EUR* Euros, *R&D* research and development

Table 2.4 Top ten pharmaceutical companies and R&D investments in 2005 and 7 years later (European Commission 2013)

Company	R&D costs (2012/2005, %)
Pfizer	−9%
Johnson & Johnson	+9%
GSK	−7%
Novartis	+69%
Sanofi	+21%
Roche	+91%
Merck & Co.	+84%
Eli Lilly	+56%
Boehringer Ingelheim	+106%
Takeda	+180%

R&D research and development

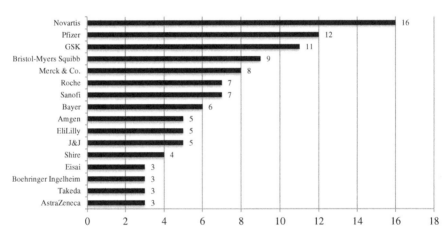

Fig. 2.2 New molecular entities (NMEs) approved by Food and Drug Administration (FDA) between 2001 and 2012 by major pharmaceutical companies (data derived from Frantz and Smith 2003; Frantz 2004, 2006; Owens 2007; Hughes 2008, 2009, 2010; Mullard 2011, 2012b, 2013, 2014a; http://www.fda.gov/Drugs/DevelopmentApprovalProcess/HowDrugsareDevelopedandApproved/DrugandBiologicApprovalReports/NDAandBLAApprovalReports/ucm373420.htm). GSK GlaxoSmithKline, J&J Johnson & Johnson

from external sources (Schuhmacher et al. 2013); 25 % of the drug candidates have been licensed and the other 25 % were acquired from outside of the companies (Schuhmacher et al. 2013). Analyzing the sources of new drugs of three of the multinational pharmaceutical companies, namely Pfizer, Roche, and Sanofi, it becomes apparent that M&A activities have played a major role in the number of new drugs launched. For example, ten NMEs have been approved by the FDA for Pfizer between 2001 and 2012. Two additional new drugs improve Pfizer's statistics directly, as two drugs had been registered for Pharmacia and Wyeth after the companies were

2 Can Innovation Still Be the Main Growth Driver of the Pharmaceutical Industry? 45

Table 2.5 Number of NMEs approved by FDA in 2001–2012 for Pfizer, Roche, and Sanofi

Company	Pfizer	Pharmacia	Wyeth	Roche	Genentech	Sanofi	Aventis	Genzyme
Year of M&A		2003	2009		2009		2004	2011
Total number of NMEs approved by the FDA (2001–2012) per single company	10	2	4	3	5	6	1	4
2001	1	1						
2002	1					1		
2003		1		1		1		
2004	1				1		1	1
2005			1					
2006	1				1			1
2007	1		1	1				
2008	1		1					1
2009						1		
2010					1	1		1
2011	1		1	1				
2012	3				2	2		
Total number of NMEs approved by the FDA (2001–2012) since acquisition of peer companies	12			6		7		
Total number of NMEs approved by the FDA (2001–2012)	16			8		11		

M&A mergers and acquisitions, *NMEs* new molecular entities, *FDA* Food and Drug Administration

acquired by Pfizer in 2003 and 2009, respectively. And four additional new drugs could be added to Pfizer as these drugs have been approved for Pharmacia or Wyeth at least 4 years before the companies have been acquired (see Table 2.4) (Table 2.5).

The multinational pharmaceutical companies listed in Fig. 2.2 have launched on average 0.6 NMEs per year between 2001 and 2012, with Novartis and Pfizer launching 1.3 and 1.16 NMEs, respectively. These figures are far below the industry goal to produce 2–3 NMEs per year per company that has been reported as a need of pharmaceutical companies to meet their growth objectives (Kola and Landis 2004;

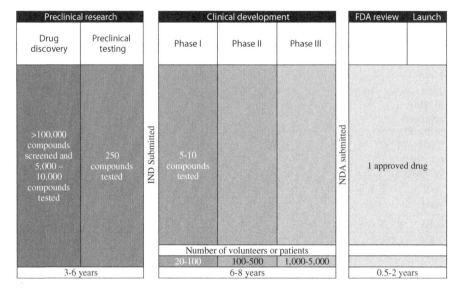

Fig. 2.3 Traditional pharmaceutical R&D process, R&D phases, and principle timelines. IND investigational new drug, NDA new drug application, FDA food and drug administration, R&D research and development

Munos 2009). Assuming a growth target of 5 % per year, a pharmaceutical company with total sales of US$ 15 billion would need to deliver 2.5–3 NMEs per year over a period of 10 years to meet this target (Kola and Landis 2004). A bigger pharmaceutical company of the size of Pfizer with total pharmaceutical sales of US$ 45 billion would need to launch 7.5–9 NMEs per year, if expecting to generate a growth of 5 % per year through pharmaceutical innovation (Kola and Landis 2004). None of the pharmaceutical companies have achieved this goal in the past years, bringing into question the dogma, that the main driver of growth in the pharmaceutical industry is innovation.

The Pharmaceutical Innovation Process

The pharmaceutical R&D process is highly regulated, lengthy, and risky. Traditionally, the process of discovering and developing a new drug is divided into preclinical research and clinical development, followed by a review and launch phase (see Fig. 2.3).

The Success Rates of Pharmaceutical R&D

As indicated in Fig. 2.3, pharmaceutical R&D has a low probability of success (PoS). Only one out of more than 100,000 compounds that have been screened in discovery research and, thereof, 10,000 compounds that have been tested during preclinical research make it to the market. In total, the probability of discovering, developing, and registering an NME has been estimated to be around 4% (Paul et al. 2010; also see 2013 CMR International Pharmaceutical R&D Factbook, http://cmr-thomsonreuters.com/pdf/fb-exec-2013.pdf). Table 2.6 summarizes some articles and highlights the probabilities per phase of drug R&D.

CMR reported for the preclinical phase, Phase I and Phase II of clinical development, success rates per phase of 67, 46, and 19%, respectively (2013 CMR International Pharmaceutical R&D Factbook, http://cmr-thomsonreuters.com/pdf/fb-exec-2013.pdf). In particular, the low PoS for the early clinical phases represents the goal that potentially unsuccessful compounds should fail early and inexpensively.

The underlying causes of the high attrition rates are manifold. Differences may depend on the drug class, the therapeutic area, the type of disease, the source of the drug candidate, and the size of the company. It has been reported that adverse pharmacokinetics and bioavailability were a major cause of attrition in the 1990s (Kola and Landis 2004). In the same opinion letter, it was stated that the lack of efficacy and safety were the major reasons for the low PoS in clinical development in 2000. In an analysis of ten big pharmaceutical companies in the period of 1991–2000, the reasons for attritions have been analyzed as being primarily efficacy and safety issues (Kola and Landis 2004).

In a review of the FDA approvals in 2012, it was reported that most of the failures in Phase II and Phase III resulted from the lack of efficacy (56%), followed by safety (28%) (Arrowsmith and Miller 2013). The lack of efficacy may be related in some therapeutic areas, such as oncology and central nervous system (CNS), with a lack of predictive animal models in the discovery research and the preclinical testing phases (Kola and Landis 2004). Today, the majority of drugs in the development refer to novel targets making drug development less predictable and, thus, less successful (Berggren et al. 2012). Biologics showed a higher PoS from Phase I to submission than small molecule drugs (SMOLs; DiMasi et al. 2010). The PoS of drugs that addressed acute diseases was also higher than the PoS of drugs treating chronic diseases (Pammolli et al. 2011). Furthermore, it could be shown that in-licensed drug candidates have a higher PoS for Phase I to submission than self-originated drugs (DiMasi et al. 2010) (Fig. 2.4). Finally, the size of a company may also have an impact on the attrition rates. While large organizations have a mean PoS of 7.86% from Phase I to submission, small organizations have a PoS of 6.07% (Pammolli et al. 2011). In the same context, biotechnology organizations seem to have lower success rates in clinical development than nonbiotechnology companies (Pammolli et al. 2011).

Further reasons for the low PoS of pharmaceutical R&D may be founded in:

- An advanced complexity of drug targets
- The higher proportion of novel drug targets
- The competition in target selection, as half of the drug targets are pursued by two or more pharmaceutical companies (Agarwal 2013)
- The complex process of target validation (Sams-Dodd 2005)

Table 2.6 Success rates per phase of pharmaceutical R&D

Period	Literature	Phase: PoS
2003	DiMasi et al. (2003)	Probability for entering phase (%) starting with Phase I: Phase I: 100.0% Phase II: 71.0% Phase III: 31.4%
2006	DiMasi JA. J Health Econ. 2006;10:107–42	Probability for entering phase (%) starting with Phase I: Phase I: 100.0% Phase II: 75.0% Phase III: 36.2%
2010	DiMasi et al. (2010)	Probability for submitting a new drug: Phase I to submission (total): 19% Phase I to submission (biologics): 32% Phase I to submission (SMOLs): 13%
2010	Paul et al. (2010)	Probability per phase: Preclinical to registration: 4.1% Target to hit: 80% Hit to lead: 75% Lead optimization: 85% Total discovery research: 51% Preclinical testing: 69% Phase I: 54% Phase II: 34% Phase III: 70% Submission to launch: 91%
2011	Pammolli et al. (2011)	Average success rates: PoS for acute diseases: 8.77% PoS for chronic diseases: 6.88% PoS of small organizations: 6.07% PoS of large organizations: 7.49% PoS of biotech: 5.14% PoS of nonbiotech: 7.86%
2012	Berggren et al. (2012)	Probability of clinical development (including review and launch): Phase I to launch (total): 8.3% Phase I to launch (SMOLs): 7% Phase I to launch (biologics): 12%
2013	2013 CMR International Pharmaceutical R&D Factbook (http://cmr.thomsonreuters.com/pdf/fb-exec-2013.pdf)	Probability per phase: Preclinical: 67% Phase I: 46% Phase II: 19% Phase III: 77% Registration: 90

SMOLS small molecule compounds, *NCEs* new chemical entities, *NBEs* new biological entities, *R&D* research and development, *PoS* probability of success. *CMR* Center for Medicine Research International

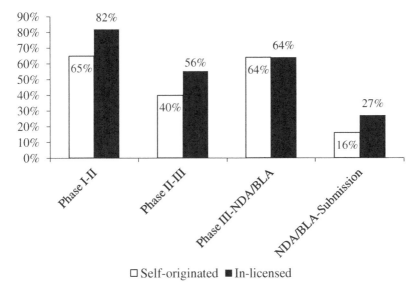

Fig. 2.4 Phase transition rates of self-originated and in-licensed drug candidates (Data derived from: DiMasi et al. 2010)

- The target-based drug discovery
- The higher demands of regulatory authorities
- A broader knowledge base increasing the complexity of clinical trials
- The greater complexities of bigger multicentric clinical trials

In an analysis of 259 drugs that were launched between 1999 and 2008, it was shown that the phenotypic screening toped the target-based approach (Swinney and Anthony 2011). Thirty-one percent of the first-in-class drugs that were analyzed were based on a phenotypic screening, while 23% were results of a target-based screening, 7% were modified natural products, and 33% were biologics. This is in comparison to the follower drugs that were analyzed, of which 51% were based on a target-based approach, 18% on the phenotypic screening, 8% on natural products, and 19% were biologics (Swinney and Anthony 2011). It was concluded that the hypothesis-driven target-based approach may contribute to higher attrition rates than the older and perhaps more productive method of drug research—the phenotypic screening (Swinney and Anthony 2011, Scannel et al. 2012). The challenge is that targets are parts of complex networks whose interactions can lead to unpredictable results. Most first-in-class drugs were discovered by phenotypic screening rather than by the target-based approach (Swinney and Anthony 2011).

The extremely low PoS of pharmaceutical R&D necessitates that pharmaceutical companies need to have an enormous number of drug projects in their R&D pipelines to ensure a continous flow of new drugs to the market.

The Timelines of Pharmaceutical R&D

The high number of R&D projects in the project portfolios of pharmaceutical companies together with the long timelines in preclinical research and drug development make drug R&D complex. Today, the total time from drug discovery to the registration of a new drug is about 14 years (Pammolli et al. 2011; Remnant et al. 2013). Table 2.7 highlights some research results and review findings in respect with R&D timing.

There have been diverse results reported in the past years on the total timing for drug R&D, which last on average between 12.3 and 14 years (Paul et al. 2010; Pammolli et al. 2011; Remnant et al. (2013); also see 2013 CMR International Pharmaceutical R&D Factbook, http://cmr-thomsonreuters.com/pdf/fb-exec-2013.pdf). The average clinical development phase and the average approval time for drugs approved between 2005 and 2009 were 6.4 years and 1.2 years, respectively (Kaitin and DiMasi 2009). It has been reported that the enactment of the PDUFA in 1992 resulted in a reduction of the average approval times by the FDA, which compensated the increasing time for the clinical development phases that have been reported in the studies listed in Table 2.7 (Kaitin and DiMasi 2009).

In a new and detailed analysis from 2010, it could be shown that discovery research, ranging from target identification to lead optimization, lasts 50 months on average, while the phases of preclinical testing and clinical development lasts for 12 and 78 months, respectively. The phase from submission to launch of a new drug lasts 18 months on average (Paul et al. 2010).

Differences in the timelines of clinical development phases of various therapeutic classes have also been reported. New drugs addressing CNS lasted longest, needing 10 years, while drugs for the treatment of AIDS antiviral had the shortest time lines, needing 4.9 years on average (Kaitin and DiMasi 2009).

In the past 5 years, the relative number of reviews by the FDA has been at a constant rate of 36–46% of all NMEs approved by the FDA. In consequence, the impact of time saving by an advanced FDA review process has been notable.

In the studies in Table 2.7, timelines for basic research and post-approval times have not been included. Assuming that basic research in respect to a drug target lasts for several years before enough knowledge is available, that is a good rational to start with applied research of pharmaceutical R&D, and assuming that the post-approval Phase IV trials continue for years, the entire process of pharmaceutical R&D lasts for at least two decades.

The Cost of Pharmaceutical R&D

The low PoS in pharmaceutical R&D together with the long timelines and the strict regulatory requirements that make drug R&D so complex, result in enormously high costs for pharmaceutical innovation. In particular, the long timelines have an enormous impact on the costs of pharmaceutical R&D. As drug costs are associ-

Table 2.7 Average timelines of pharmaceutical R&D phases

Period	Literature	Time/phase
1991	DiMasi (1991)	Average clinical phase lengths for approved NCEs: Phase I: 14 months Phase II: 25.9 months Phase III: 36.8 months
2003	Reichert JM. Nat Rev Drug Discov. 2003;2:695–702	Average time of Phase I to approval: 6–8 years Mean phase lengths of clinical development and approval (1982–2001): Anti-infective: 74.5 months Antineoplastic: 116.0 months Cardiovascular: 103.3 months Endocrine: 115.3 months Immunological: 100.2 months
2003	DiMasi et al. (2003)	Time from start of clinical testing: Phase I to submission: 72.1 months Phase I to marketing approval: 90.3 months Average phase times for investigational compounds (1985–2000): Phase I: 21.6 months Phase II: 25.7 months Phase III: 30.5 months
2009	Kaitin and DiMasi (2009)	Mean clinical phase times (Phase I to submission) and mean approval times of NCEs and NBEs approved between 1980 and 2009: 1980–1984: 5.7 years/2.8 years 1985–1989: 5.8 years/2.7 years 1990–1994: 6.4 years/2.4 years 1995–1999: 6.5 years/1.4 years 2000–2004: 6.6 years/1.5 years 2005–2009: 6.4 years/1.2 years
2010	Paul et al. (2010)	Average time from preclinical to registration: 13.5 years Average time per phase: Target to hit: 12 months Hit to lead: 18 months Lead optimization: 24 months Preclinical testing: 12 months Phase I: 18 months Phase II: 30 months Phase III: 30 months Submission to launch: 18 months
2011	Pammolli et al. (2011)	Average time for clinical development to submission increased from 9.7 years for new drugs launched in the 1990 to 13.9 years for new drugs launched between 2000 and 2008
2013	2013 CMR International Pharmaceutical R&D Factbook (http://cmr-thomsonreuters.com/pdf/fb-exec-2013.pdf)	Average time from preclinical to registration: 12.3 years

Table 2.7 (continued)

Period	Literature	Time/phase
2013	Remnant et al. (2013)	The total time for drug R&D: 14 years

R&D research and development, *NCEs* new chemical entities, *NBEs* new biological entities, *SMOL* small molecule compounds, *CMR* Center for Medicine Research International

ated with R&D expenditures that were invested many years ago, drug costs need to be capitalized until the day of return on investment. Excluding any other factors and assuming today's timelines of 14 years for drug R&D, the total R&D costs of US$ 1.8 billion and total sales of US$ 250 million in the first year, followed by US$ 500 million in the second year, and US$ 1000 million in the third year, an additional 3 years are required till the day of return on investment. As a consequence, the costs of drug R&D need to be capitalized on a period of 14 years plus an additional time of 3 years. An increase in the interest rate and any prolongation of the R&D timelines has a negative impact on costs of drug R&D. The table 2.8 summarizes the development of costs of drug R&D in the past years.

It has been reported that the R&D costs have doubled every 8.5 years since 1950 (Munos 2009). The annual increase in capitalized costs per NME has been calculated to be 12.3%. (Munos 2009). Before the 1990s, costs for drug R&D had been less than US$ 250 million (DiMasi 1991). In 2003, the average out-of-the-pocket costs were already US$ 403 million, and the capitalized costs had been calculated to be US$ 802 million (DiMasi et al. 2003). It has been stated that the increase was primarily related to increasing costs in clinical development (+350% from 1991 to 2003) (DiMasi et al. 2003). Today, the total out-of-the-pocket costs for drug R&D have been calculated to be US$ 873 million, while the total capitalized costs are US$ 1.778 billion (Paul et al. 2010). It has been reported that the clinical development phases from Phase I to submission account for 63% of these total R&D costs (Paul et al. 2010).

The reasons for the increasing R&D costs may relate with:

- New technologies in drug research, such as combinatorial chemistry, DNA sequencing, high throughput screening, and computational drug design, that have been implemented to increase the throughput.
- The increasing clinical trial sizes
- The increasing costs for clinical infrastructure
- A greater complexity of clinical trials conducted for drugs to treat chronic diseases (DiMasi et al. 2003)
- A higher number of R&D personnel (Cohen 2005).
- In particular, the clinical development functions accounted for more than 50% of all R&D expenditures.

The cost calculations and assumptions listed in Table 2.8 may not be complete, as they do not include costs for basic research, costs related with Phase IV trials, costs for regulatory approvals in non-US markets, or costs for developing the same drug in new indications. Whereas the high capitalized costs are due to the long R&D

Table 2.8 Costs of pharmaceutical R&D and costs per phase of R&D

Period	R&D costs	Literature
1950–1960	US$ 0.5 million (data derived from: DiMasi 1991)	Schnee JE. Development costs: determinants and overruns. J Bus. 1972;347–374
1976	US$ 54 million[57]	Hansen RW. Pharmaceutical development costs by therapeutic categories, University of Rochester Graduate School of Management Working Paper No. GPB-80–6. 1980
1987	US$ 231 million	DiMasi (1991)
2003	US$ 802 million	DiMasi et al. (2003)
2007	US$ 1318 million	DiMasi JA, Grabowski HG. Managerial Decis Econ. 2007;28: 469–79
2010	US$ 1778 million	Paul et al. (2010)
2013	US$ 1219 million	Remnant et al. (2013)

USD US Dollars, *R&D* research and development

timelines, most of the out-of-the-pocket costs are associated with the low PoS of drug R&D and, thus, with the costs of failed research projects and development compounds (Paul et al. 2010; Scannell et al. 2012).

A Steady-State R&D Model

It has been reported that 24 research projects need to be started every year to statistically yield in one new drug launched annually (Paul et al. 2010). In view of the pharmaceutical companies' growth objectives and goals to produce two to three NMEs per year, pharmaceutical companies would need to start more than 60 research projects in the phase target to hit annually.

Table 2.9 summarizes the idealistic situation of an R&D project portfolio of a pharmaceutical company that is launching 2.5 NMEs into the market every year. Statistically, the company would need to start more than 60 research projects annually, if doing internal R&D only, to have a steady state of 32 projects in Phase I, 28.8 projects in Phase II, and 9.8 projects in Phase III. Given the data of Table 2.9, it is obvious that multinational pharmaceutical companies need to have a certain R&D size of more than 100 active projects in clinical development phases to be successful.

The Reduced R&D Efficiency

R&D efficiency has been defined as the ability of an R&D organization to translate an input, such as the investment, into an output, such as the number of new products launched to the market (Paul et al. 2010). Scannell and coauthors have analyzed the

Table 2.9 Fictive R&D pipeline required to statistically provide 2.5 NMEs/per year. P(TS) (probability of technical success) and Timing from Paul et al. (2010)

	Target to hit	Hit to lead	Lead optimization	Pre-clinical	Phase I	Phase II	Phase III	Submission to launch	New drugs
p(TS) (%)	80	75	85	69	54	34	70	91	
Timing	1.0	1.5	2.0	1.0	1.5	2.5	2.5	1.5	
Projects needed for 2.5 launches per year	60.6	72.7	72.7	30.9	32	28.8	9.8	4.1	2.5

decline of the pharmaceutical R&D efficiency in a period from 1950 to 2010 and concluded that the number of new drugs approved per US$ 1 billion halved nearly every 9 years in the past 60 years, reaching a level of US$ 1 billion for 1 NME in 2000 (Munos 2009; Scannell et al. 2012). This trend is the result of a development in the pharmaceutical industry, whereby the number of new drugs launched by the industry was constant while the costs per new drug increased steadily. Today, the capitalized costs per new drug have been calculated to be US$ 1.778 billion although it could be assumed that the actual full costs of drug R&D are even higher (Paul et al. 2010). In an analysis by PriceWaterhouseCoopers (PWC), the costs per launch of a new drug were analyzed as the ratio of the total R&D costs of the industry to the total number of new drugs approved by the FDA (PWC 2012). It came out that the costs per NME in the years 2002–2011 were up to 4.6 billion (Fig. 2.5).

In a series of three papers, Forbes analyzed the costs of inventing new drugs, concluding that the average costs of drug development of top pharmaceutical companies are between US$ 3.3 and 13 billion (Harper 2012a, b, 2013). It has also been concluded that smaller pharmaceutical companies need less money to launch a new drug. This may relate to the fact that only successful small companies have been considered in the statistic and failed companies were disregarded. Finally, it was investigated that the top pharmaceutical companies, that have launched more than four NMEs in the 10 years from 2002 to 2011, invested more than US$ 5 billion per new drug. Table 2.10 summarizes an analysis of 11 multinational pharmaceutical companies, including their R&D costs, the number of NMEs approved by FDA between 2001 and 2012, and their R&D efficiencies.

In the analysis of Table 2.10, it becomes apparent that the pharmaceutical companies listed had total R&D costs of US$ 4.5–18.6 billion per new drug approved by FDA in the past 10 years. On average, pharmaceutical companies invested US$ 9 billion per new drug (median US$ 7.6 billion), an amount that is significantly higher than the figures that have been calculated in previous publications (see Table 2.8).

2 Can Innovation Still Be the Main Growth Driver of the Pharmaceutical Industry? 55

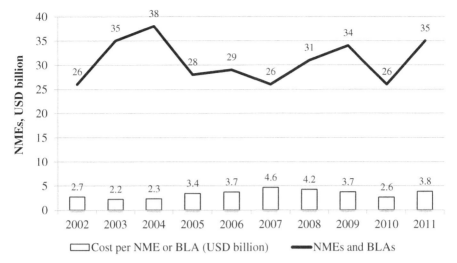

Fig. 2.5 Development of industry-wide total R&D costs per new drug. R&D research and development (Data derived from: PWC 2012)

Table 2.10 R&D efficiencies (2001–2012) of multinational pharmaceutical companies

	Total number NMEs (2001–2012)	Total R&D costs (USD million, 2001–2012)	R&D efficiency (costs per launch)
AstraZeneca	3	55,959	18,653
Roche	7	83,888	11,984
Takeda[a]	2	23,376	11,688
Sanofi[b]	4	38,912	9728
EliLilly	5	47,949	9590
Pfizer[c]	12	91,367	7614
GSK	11	76,538	6958
Boehringer Ingelheim[d]	3	20,727	6909
Amgen	5	34,119	6824
Novartis	16	82,004	5125
Bristol-Myers Squibb	9	40,292	4477

[a] Data of Takeda from 2006 to 2012
[b] Data of Sanofi from 2005 to 2012
[c] Data of Pfizer from 2002 to 2012
[d] Data of Boehringer Ingelheim from 2004 to 2012
NMEs new molecular entities, *R&D* research and development, *USD* US Dollars

Possible reasons for the low R&D efficiency have been discussed previously in context of the low PoS of pharmaceutical R&D and the high costs for pharmaceutical innovation. Furthermore, an insufficient number of projects in preclinical and early clinical phases may have negatively impacted the R&D efficiency (Paul et al. 2010). The increasing number of approved drugs raised the hurdle for approval and reimbursement of new drugs (Scannell et al. 2012). In the same way, a lower risk tolerance of drug regulators may have increased both the challenges for launching new drugs and the development-associated costs (Scannell et al. 2012). It has been reported that the target-based screening in drug discovery replaced the phenotypic screening and that the potential of drug-screening methods in discovery research and their impact on timelines and PoS have been overestimated, while costs were increased (Swinney and Anthony 2011; Scannell et al. 2012). In the same context, a general belief that high-affinity binding to a single biological target is directly linked to a disease and, thus, the activity or inhibition of that target results in a medical benefit might be incorrect and misleading (Scannell et al. 2012). In addition, an increasing number of mergers might have influenced the efficiency of pharmaceutical R&D negatively (LaMattina 2011). And, finally, it has been said that the low-hanging fruits have already been picked, resulting in technically more complex investigations for new drug targets and related preclinical and clinical studies (Scannell et al. 2012).

PhRMA reported a stagnating overall R&D expenditure for its members since 2007 (PhRMA 2013) of minimum US$ 46.4 billion (2009) and maximum US$ 50.7 billion (2010). The industries' output, measured in the total number of NMEs per year, has also been at a constant level during this time period. Both indicators show that, at least for the past years, the R&D efficiency of the pharmaceutical industry has not been reduced further and, in view of the NME output in the years 2011 and 2012 with 30 and 39 NMEs, respectively, there is hope for an increase in R&D efficiency in the future (see Table 2.11).

Table 2.11 Overall R&D efficiency of the pharmaceutical industry in the years 2007–2012

Year	Total number of drugs (NMEs) approved by the FDA	Total R&D expenditures of PhRMA members (USD billion)	Cost per new drug (USD billion)
2007	18	47.9	2.66
2008	24	47.4	1.98
2009	25	46.4	1.86
2010	21	50.7	2.41
2011	30	48.6	1.62
2012	39	48.5	1.24

NMEs new molecular entities, *FDA* Food and Drug administration, *PhRMA* Pharmaceutical Research and Manufacturers of America, *USD* US Dollars

Consequences of the Reduced R&D Efficiency

The results of the reduced R&D efficiency have been enormous for the industry, as pharmaceutical innovations have been up to this point the major driver of its growth. The consequences have been either the attempt to reduce R&D costs, attrition rates and cycle times, or pharmaceutical companies have attempted to increase the R&D productivity, being defined as the relationship between the commercial value created by a new medicine, and the investment required to generate that new medicine (Paul et al. 2010). In detail, an increase in R&D productivity is possible by influencing the elements with the greatest impact on productivity, namely by an increase of the number of projects in the R&D pipeline, or an increase of the probability of technical and regulatory success per pipeline project, or an increase of the (financial) value per project, or a reduction of the cycle times, or the reduction of the costs per pipeline project. It has been reported that an improvement in R&D efficiency and R&D productivity is, in particular, possible by reducing attrition rate in Phase II and Phase III of clinical development (Paul et al. 2010).

Increasing the Number of Projects in the R&D Pipeline

The global R&D pipeline, which is the number of pipeline projects in the phases of preclinical testing to market launch, has increased enormously in the past years. Since 2001, the total number of projects listed in the pipelines of pharmaceutical companies worldwide has increased from 5995 to 11,307 (Citeline 2013). In 2014, 5484 projects were listed in the preclinical testing phase, 1541 in Phase I, 2011 in Phase II, 744 in Phase III, 170 in a preregistration phase and 1074 in market launch (Citeline 2013). In the past 3 years, the global pipeline increased in all phases of clinical development resulting in corporate R&D pipelines of top pharmaceutical companies of more than 200 pipeline projects. The group of companies with the most projects in their R&D pipeline is: GSK (261), Roche (248), Novartis (223), Pfizer (205), and AstraZeneca (197) (Citeline 2013). Within the top 25 pharmaceutical companies with the most projects in their R&D pipeline, 17 companies increased their pipeline size between 2013 and 2014. At the same time, the number of companies with an active R&D pipeline increased from 2745 (2013) to 3107 (2014), giving a reasonable expectation that the global pipeline size will also increase in the future, if enough venture capital is allocated to early drug research.

Reducing Costs of R&D

The total R&D expenditures increased enormously from 1995 (US$ 15.2 billion) to 2007 (US$ 47.9 billion; PhRMA 2013). Since 2007, the members of the PhRMA have reduced their financial efforts in R&D and total numbers are stagnating. Today, the total R&D investments are US$ 48.5 billion (see Fig. 2.6).

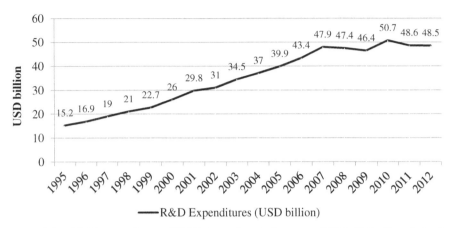

Fig. 2.6 Total R&D expenditures of PhRMA members in the years of 1995–2012. (Data derived from PhRMA 2013)

The nominal expenditures of the top pharmaceutical companies are listed in Table 2.12.

The nominal R&D spending of 12 of the top pharmaceutical companies has been developing differently in the past years. Between 2007 and 2013, eight pharmaceutical companies have increased their total R&D spending against the industry trend, while four of the companies have decreased their R&D costs. At least for this group of companies, there is no clear strategic trend towards reduced nominal R&D expenditures, although some companies have cut their R&D efforts enormously. As for example, Pfizer announced in 2011 to reduce R&D costs by closing labs and reducing research spending by up to US$ 3 billion[1] GSK published in 2012 to reduce the R&D and manufacturing organizations by 2016.[2] And Merck & Co. announced a 17% reduction in R&D personnel.[3]

Generally, a reduction in R&D costs is related to:

- A general reduction of R&D personnel, with a focus on lowering costs by reducing overhead functions in R&D, which is typically more prevalent in bigger organizations.
- A greater focus in project and portfolio management on project costs (David et al. 2010).
- Outsourcing to low-cost countries to reduce operational and infrastructure costs (David et al. 2010).

---

[1] http://www.bloomberg.com/news/2011-02-01/pfizer-fourth-quarter-net-topss-analyst-estimates-shares-fall-on-outlook.html

[2] http://www.pharmatimes.com/article/13-02-07/GSK_puts_faith_in_pipeline_and_cuts_costs_after_tough_2012.aspx

[3] http://www.fiercepharma.com/story/skinny-earnings-cost-cuts-boost-merck-bristol-myers-forest-fx-hits-sanofi/2014-04-29

Table 2.12 Nominal R&D spending of top pharmaceutical companies (2007–2013)

	Amgen (USD billion)	AstraZeneca (USD billion)	BI (USD billion)	BMS (USD billion)	Eli Lilly (USD billion)	GSK (BP billion)	MSD (USD billion)	Novartis (USD billion)	Pfizer (USD billion)	Roche (CHF billion)	Sanofi (EUR billion)	Takeda (JPY million)
2007	3.266	5.089	1.730	3.282	3.487	3.327	4.883	6.400	8.089	8.385	4.537	193.301
2008	3.030	5.179	2.109	3.512		3.681	4.805	7.200	7.945	8.845	4.150	275.788
2009	2.864	4.409	2.215	3.647	4.327	4.106	5.845	7.300	7.845	9.874	4.091	453.056
2010	2.984	5.318	2.453	3.566	4.884	4.457	11.111	8.100	9.413	9.050	3.884	296.392
2011	3.167	5.523	2.516	3.839	5.021	3.687	8.467	9.200	8.681	8.073	4.101	288.874
2012	3.380	5.243	2.795	3.904	5.278	3.979	8.168	9.100	7.482	8.475	4.905	281.885
2013	4.083	4.821	2.743	3.731	5.531	3.923	7.503	9.600	6.678	8.700	4.770	324.292
Relative change 2007–2013 (%)	+25	−6.3	+59	+13.6	+59	+17.9	+54	+50	−17.5	−3.8	−5.1	+68

EUR = Euros, *BP* = British Pounds, *USD* = US Dollars, *JPY* = Japanese Yen, *CHF* = Swiss Franks, *BI* = Boehringer Ingelheim, *BMS* = Bristol-Myers-Squibb, *GSK* = GlaxoSmithKline, *MSD* = Merck & Co.

In an analysis on strategic outsourcing, CEPTON Strategies reported a 15% share of outsourcing for R&D with a total outsourcing volume of US$ 70 billion in 2008.[4] Today, the top clinical research organizations (CROs) are the full-service providers Quintiles and Covance with total revenues in 2013 of US$ 3.8 and 2.4 billion, respectively.

Some pharmaceutical companies have tried to use the M&As of the past years to generate nominally bigger R&D organizations with larger project portfolios, higher cross-fertilization, better economies of scale, and reduced R&D rates. As for example, Pfizer has been through two mega-mergers in the past 10 years producing a company that finally did not grow in the number of employees or in the relative rate of R&D investment, but increased the nominal spending in R&D and its portfolio size.

In addition to Pfizer (12.9%), GSK and Sanofi have been able to reduce their R&D rates in the past years and have achieved an R&D rate in 2013 of 14.3 and 14.5%, respectively. All three figures are clearly under the historical industry benchmark of 20%, showing a newer industry trend towards significant lower R&D investments. These figures are still far from the R&D rate of the worldwide biggest generic company Teva that had total sales in 2013 of US$ 20,314 billion and total R&D costs of US$ 1.422 billion with a resulting R&D rate of 7%. However, the figures of Teva show the theoretically possible savings for some of the multinational pharmaceutical companies on their way to reduce R&D costs, if they decide to change their business model from purely R&D-based to a generic-based pharmaceutical company.

Another option to increase the R&D efficiency has been the change in the R&D business model from a centralized in-house R&D to smaller, more focused, and better manageable R&D units (Garnier 2008). Thereby, the archetype has been the biotechnology industry and the reorganizations that took place in the past years aimed at providing a more biotech-like and entrepreneurial spirit in pharmaceutical R&D organizations (Douglas et al. 2010; Zhong and Mosley 2010).

Measuring Performance and Managing the Project Portfolio Actively

A greater management attention towards project costs, resource allocation, and the active management of the project portfolio has been described as an effective method and success factor. The R&D pipeline size and the progress of R&D projects should be managed in accordance with a steady-state pipeline model. In view of the companies' success rates per phase and the timing of the projects, a model needs to be set up that enables pharmaceutical companies to continuously deliver

---

[4] CEPTON Strategies – Strategic outsourcing across the pharmaceutical value chain (http://www.cepton.net/publications/download/cepton-Strategic-outsourcing-across-the-pharmaceuticals-value-chain.pdf)

new drugs to the market. The focus of R&D needs to change from late-stage development projects that may provide success in the near term to all phases of drug R&D. Consequently, an adequate number of projects need to be in all preclinical research phases, followed by a sufficient number of projects in all phases of clinical development. As the financial and human resources of pharmaceutical companies are limited, the number of projects in the late-stage development needs to be reduced to a level that enable the company to reallocate the free resources to earlier phases, in particular to Phases I and II, to finally increase the success rate in a continuous pipeline model. To run a portfolio model, R&D performance metrics need to be installed, including the access to benchmark data of competitor companies. The portfolio decisions need to be based on medical need, technical feasibility, and commercial value. The critical path of each R&D project needs to be identified, and project management along the critical path needs to be optimized to reduce cycle times. Finally, pharmaceutical companies need to invest only in R&D tasks that support project-related decision making, reduce costs of technology development, and, thus, free up resources that can be allocated to drug projects. All efforts together should help to focus on those R&D tasks that are related to high-priority-drug R&D projects, reduced cycle times, and reduced attrition rates of drug projects.

Opening R&D Towards External Innovation

In view of the increased pressure on time and costs of pharmaceutical R&D, pharmaceutical companies needed to enlarge their portfolio breadth to meet at least parts of their growth objective by launching new drugs. Today, pharmaceutical companies use open innovation to harness innovation externally (Chesbrough 2003; Hunter and Stephens 2010). For example, companies fill their internally generated project portfolios by acquiring drug candidates (see Table 2.13). It has been described that multinational pharmaceutical companies have acquired on average 50% of their pipeline projects from external sources (Schuhmacher et al. 2013).

Parallel to the development of project portfolios that were generated from internal and external sources, some pharmaceutical companies have aligned their organizational structures to access external innovation more efficiently. For example, GSK launched its Center for Excellence for External Drug Discovery in 2007, an externally focused R&D center that facilitates drug discovery alliances with external partners.[5] In 2010, Pfizer established the Centers for Therapeutic Innovation (CTI), an open innovation model that aims at founding global partnerships between Pfizer and academic medical centers.[6] Additionally, as early as 2002, Eli Lilly started the Fully Integrated Pharma Network (FIPNet), the Phenotypic Drug Discovery Initiative, the Target Drug Discovery Initiative, and Chorus, (Ernst & Young 2010).[7,8] Further examples of open innovation initiatives are the crowd-sourcing

---

[5] http://www.out-sourcing-pharma.com/Preclinical-Research/GSK-opens-Centre-of-Excellence

[6] http://www.pfizer.com/research/rd_works/centers_for_therapeutic_in- novation.jsp

[7] https://openinnovation.lilly.com/dd/

[8] http://www.choruspharma.com/about-us.html

Table 2.13 Key R&D pipeline figures of multinational pharmaceutical companies. (Data derived from EvaluatePharma® 2011)

	Total number of R&D projects	Number of organic R&D projects	Number of R&D projects accessed by company acquisition	Number of R&D projects licensed	Total externally sourced R&D pipeline (%)	Externally sourced R&D pipeline by licensing (%)
Amgen	62	30	24	8	52	13
Astra Zeneca	102	44	30	28	57	27
Boehringer Ingelheim	56	46	0	10	18	18
BMS	103	42	42	19	59	18
Eli Lilly	111	79	15	17	29	15
GSK	241	136	22	83	44	34
Merck & Co.	113	50	34	27	56	24
Novartis	176	104	28	44	41	25
Pfizer	143	75	53	15	48	10
Roche	143	74	47	33	48	23
Sanofi	116	33	39	54	72	47
Shire	19	19	NA	NA	80	NA
Takeda	65	40	19	32	38	49

NA not applicable, *R&D* research and development

platform Grants for Targets of Bayer and PD2 of Eli Lilly (Lessl et al. 2011, see footnote 7). The potential of Open Source Drug Discovery and the African Network for Drug and Diagnostics Innovation have also been discussed (Munos 2010).

In the course of the opening of the R&D organizations, collaborations with academic institutions have also played an important role. It has been published that 30% of all novel drugs come from academia and that academic institutes are a major source of drug projects (Kneller 2010).

In addition to drug targets, knowledge, and know-how in some therapeutic areas, academic collaboration partners can provide technologies and capabilities that are of value for pharmaceutical companies. As for example, the Division of Signal Transduction Therapy (DSTT) is a collaboration between the University of Dundee, the Medical Research Council (MRC), and six pharmaceutical companies, namely AstraZeneca, Boehringer Ingelheim, GSK, Janssen Pharmaceutica NV, Merck Serono, and Pfizer to perform research on the development of new drug treatments for major global diseases.[9] The Tuberculosis Drug Accelerator (TBDA) is a consortium of Abbott, AstraZeneca, Bayer, Eli Lilly, GSK, Merck & Co., and Sanofi together

---

[9] http://app.dundee.ac.uk/pressreleases/2012/may12/drugdiscovery.htm

2   Can Innovation Still Be the Main Growth Driver of the Pharmaceutical Industry?   63

Table 2.14 Major collaborations between academic institutes and pharmaceutical companies in 2012. (Data derived from http://www.fiercebiotech.com/slideshows/20-major-pharma-academic-alliances-2012)

Year	Pharmaceutical company	Academic partner	Scope of the collaboration
2012	Sanofi	UCSF	New treatments for type I and type II diabetes
2012	Pfizer, Eli Lilly, and Merck & Co. in a consortium called Asian Cancer Research Group (ACRG)	University of Singapore, The University of Hong Kong	Analyzing cancers impacting Asian populations
2012	Novi Nordirsk	University of Oxford	Biomarker development
2012	UCB	University of Oxford	New immunology and neurology medications
2012	BMS	Vanderbilt University	New treatments for Parkinson's disease
2012	Novartis	University of Pennsylvania	Research on personalized T cell therapies for the treatment of cancer

UCB Union chimique belge, *BMS* Bristol-Myers Squibb, *UCSF* University of California, San Francisco, *MSD* Merck & Co.

with Texas A&M University, Weill Cornell Medical College, and the Bill & Melinda Gates Foundation that was created to discover new treatments against tuberculosis.[10] Further major pharma–academia collaborations are compiled in Table 2.14.

Sustainable Growth in Times of Reduced R&D Efficiency

If R&D efficiency, defined as the costs per launch, is reduced, and if pharmaceutical companies are not able to increase their R&D efficiencies by the measures discussed before, they still can try to compensate it by increasing the value per drug launched, if the payers are willing to pay high prices for the new drugs. If the increase in the value per drug compensates for the rising costs completely, the R&D productivity is stable. If it overcompensates, the R&D productivity would increase. It has been highlighted that the value of one NME, measured as the 5-year post-launch sales, grew in the time period from 2010 (US$ 10 billion) to 2012 (US$ 16 billion) (EvaluatePharma 2013b). In contrast to this analysis, it is expected that the average peak sales per NME declines from US$ 900 million (2012) to US$ 600 million (2015), showing the increasing difficulty of offering benefits over existing treatments in

---

[10] http://www.abbott.com/news-media/press-releases/seven-pharmaceutical-companies-join-academic-researchers-to-speed-tb-drug-discovery.htm

light of the increasing price pressure (Berggren et al. 2012). In this context, the projected revenues of all NMEs launched between 2012 and 2016 are expected to be US$ 58 billion, whereas losses by patent expirations between 2013 and 2016 are forecasted to be US$ 123 billion, showing that the new revenues will not compensate for the revenue losses by patent expirations in the industry (Berggren et al. 2012; Schacht 2012; EvaluatePharma 2013a). It is challenging to project whether the industry will compensate the decline of R&D efficiency with an increase in value per drug launched, but the numbers presented herein show at least that the pharmaceutical industry needs to invent alternative scenarios to maintain sustainability.

Increasing Pressure from Generic Drugs

Pharmaceutical innovation has been, until now, the major driver of growth for the pharmaceutical industry. The reduced R&D efficiency and the challenges in increasing the value per drug launched make it necessary that pharmaceutical companies keep an eye out for other growth options. In a 2010 forecast by KPMG, it was said that growth of NMEs in the period of 2010–2015 are compensated by the losses resulting from patent expirations. Growth in the industry will come from the generics business (+US$ 47 billion) and emerging markets (+US$ 150 billion) (KPMG 2011). The total global spending on medicines has been forecasted to reach approximately US$ 1200 billion in 2017, an increase of US$ 205–235 billion from 2012.[11] In the same analysis, it has been said that growth in the developed countries will primarily come from new treatments in chronic diseases, such as cancer and diabetes. Growth in the "pharmerging" countries will result from an increase in sales in traditional therapy areas, although populations in "pharmerging" countries will also become older and obese, resulting in further growth options for the pharmaceutical industry. The worldwide prescription drug sales are forecasted to a total volume of US$ 895 billion in 2018 with a compound annual growth rate (CAGR) of 3.8 % between 2012 and 2018 (EvaluatePharma® 2013). Reviewing the growth of the global pharmaceutical market in geographical regions, two independent analyses have been made showing that the emerging countries will be the major drivers of growth with forecasted market potentials of US$ 499 and US$ 487 billion by 2020 (KPMG 2011; PWC 2012).

The challenge for the pharmaceutical industry is the low pharmaceutical sales per capita in "pharmerging" countries, which is 5–20 times lower than the pharmaceutical sales per person in developed countries (see Table 2.15). In particular, in the "pharmerging" countries, both health-care systems and private patients struggle to pay for new medicine.

---

[11] IMS Institute, The global Use of Medicine: Outlook Through 2017, http://www.imshealth.com/deployedfiles/imshealth/Global/Content/Corporate/IMS%20Health%20Institute/Reports/Global_Use_of_Meds_Outlook_2017/IIHI_Global_Use_of_Meds_Report_2013.pdf

Table 2.15 Pharmaceutical sales in selected countries in 2011. (Data derived from International Federation of Pharmaceutical Manufacturers & Associations, The Pharmaceutical Industry and Global Health, Facts and Figures 2012, http://www.ifpma.org/fileadmin/content/Publication/2013/IFPMA_-_Facts_And_Figures_2012_LowResSinglePage.pdf)

Country	Pharmaceutical sales per capita (USD)
Brazil	146
Russia	145
India	13
China	50
USA	1077
Germany	671
Japan	1007

USD US Dollars

Today, the market share of generic drugs in "pharmerging" countries is dominant. In 2012, generic drugs had a market share in China of 76%, while off-patent drugs and innovative drugs with patent protection had a stake of 20 and 4%, respectively (IMAP 2012). IMS has forecasted that generics will achieve a larger market share in developed and "pharmerging" countries by 2017 (see footnote 11). Consequently, some of the multinational pharmaceutical companies already generate today a major part of their total revenues outside the traditional main markets of Europe, USA, and Japan by selling generic drugs (PWC 2012). It has been forecasted that the emerging countries will contribute as much to global pharmaceutical profits as the USA by 2020 (KPMG 2011). Thus, even if there is an increase in the worldwide total sales of the pharmaceutical industry, the lower profits of the "pharmerging" countries result in lower profit margins of pharmaceutical companies. This development will result in lower investments in R&D in the future and will increase the pressure on R&D organizations to improve their R&D efficiencies.

Sustainability Must Come from R&D

In view of the limited growth options that are offered to the pharmaceutical sector in the coming years, pharmaceutical companies need to focus on the increase in R&D efficiency and R&D productivity. In addition to what has been said before, pharmaceutical companies should follow the following strategies:

- Focus on therapeutic areas and drug candidates with the greatest PoS.
- Focus R&D activities on drug candidates that can provide benefit to real patients' needs.
- Provide real differentiated new products.
- Focus on personalized medicine, as biomarker-based patient stratification has been cited to increase PoS across all phases in drug development of oncology drugs (Hayashi et al. 2013).

- Further, reduce R&D costs by focusing R&D on core competences and outsource nondifferentiating activities to external experts.
- Create asset pools and combine R&D activities of pharmaceutical companies.
- Provide tailor-made products for developed and "pharmerging" countries and differentiate the drug prices, respectively.

Furthermore, a mega-fund has been proposed to increase financial funding of industry-wide R&D activities, as smaller companies are critically important for discovering innovative drugs (Kneller 2010; Fernandez et al. 2012). The mega-fund could finance target identification and validation. In combination with a broader externalization of pharmaceutical R&D to smaller and specified companies, this would help to mitigate technical risks associated with early drug research, while using the competences of pharmaceutical companies in preclinical testing and clinical development (Mullard 2012a). There is hope in respect to the large number of novel targets that, if investigated and clinically validated, could be basis for new, more efficacious, and safer drugs (Berggren et al. (2012); Scannell et al. 2012). There is reasonable expectation that new drugs can provide a therapeutic benefit that comes from interacting with different targets.

Other options to increase R&D efficiency and productivity are drug repositioning and incremental innovations, such as the screening of abandoned, failed, or approved drugs for new uses, or the improvement of formulations or new uses of existing drugs (Cohen 2005; Mullard 2011). These strategies are not new for the industry, but could get more significance, if pharmaceutical companies realize that there are defense strategies other than patent rights based on novelty and inventive step. Optionally, a prolongation of the 5-year supplementary protection certificate (SPC) for pharmaceuticals might also provide more funding to for pharmaceutical R&D.

Pharmaceutical R&D is and will be a very expensive adventure with an overall low PoS and long timelines. In particular, the challenge of high costs makes it more and more difficult to pharmaceutical companies to afford R&D and to provide new drugs to the market. Any option that might increase funding, in particular, in the research of new drugs, would be very helpful and supportive and would help the pharmaceutical industry to keep sustainability.

References

Agarwal P. Novelty in the target landscape of the pharmaceutical industry. Nat Rev Drug Discov. 2013;12:575–6.

Arrowsmith J, Miller P. Trial watch: phase II and phase III attrition rates 2011–2012. Nat Rev Drug Discov. 2013;12:569.

Berggren R, et al. Outlook for the next 5 years in drug innovation. Nat Rev Drug Discov. 2012;11:435–6.

Chesbrough H. Open innovation. In The new imperative for creating and profiting from technology. Boston: Harvard Business School Press; 2003.

Citeline. Pharma R & D annual review. 2013. http://www.citeline.com/wpcontent/uploads/Annual-Review-2014b.pdf. (2013). Accessed 28 Oct 2014.

Cohen FJ. Macro trends in pharmaceutical innovation. Nat Rev Drug Discov. 2005;4:78.

David E, et al. Pharmaceutical R&D: the road to positive returns. Nat Rev Drug Discov. 2010;8:609–10.

DiMasi JA. Cost of innovation in the pharmaceutical industry. J Health Econ. 1991;10:107–42.

DiMasi JA, et al. The price of innovation: new estimates of drug development costs. J Health Econ. 2003;22:151–85.

DiMasi JA, et al. Trends in risks associated with new drug development: success rates for investigational drugs. Clinic Pharmacol Ther. 2010;87(3):272–7.

Douglas FL, et al. The case for entrepreneurship in R&D in the pharmaceutical industry. Nat Rev Drug Discov. 2010;9:683–9.

Ernst & Young. Beyond the Borders—Global Biotechnology Report 2010. 2010

European Commission—Joint Research Centre. The 2013 EU Industrial R & D Investment Scoreboard 2013. http://iri.jrc.ec.europa.eu/documents/10180/1960e4e9–37ea-4774-a8d1-c4b1629e7ab1. Accessed 28 Oct 2014.

EvaluatePharma®. Annual company reports (2011). Datamonitor® 2011.

Evaluate Pharma. 2012 Year in review. 2013a. http://info.evaluatepharma.com/rs/evaluatepharmaltd/images/EPV_Review_2012.pdf. Accessed 28 Oct 2014.

EvaluatePharma®. World Preview 2013, Outlook 2018 Returning to Growth. 2013b. http://download.bioon.com.cn/view/upload/201307/26155533_4502.pdf. Accessed 28 Oct 2014.

Fernandez J-M, et al. Commercializing biomedical research through securitization techniques. Nat Biotechnol. 2012;30:964–75.

Frantz S. 2003 approvals: a year of innovation and upward trends. Nat Rev Drug Discov. 2004;3:103–5.

Frantz S. 2005 approvals: safety first. Nat Rev Drug Discov. 2006;5:92–3.

Frantz S, Smith A. New drug approvals for 2002. Nat Rev Drug Discov. 2003;2:95–6.

Frye S, et al. US academic drug discovery. Nat Rev Drug Discov. 2011;10:409–10.

Garnier JP. Rebuilding the R & D engine in big pharma. Hav Bus Rev. 2008;86:68–79.

Harper M. The cost of creating a new drug now $ 5 billion, pushing big pharma to change. 2012a. http://www.forbes.com/sites/matthewherper/2013/08/11/how-the-staggering-cost-of-inventing-new-drugs-is-shaping-the-future-of-medicine/. Accessed 28 Oct 2014.

Harper M. The truly staggering cost of inventing new drugs. 2012b. http://www.forbes.com/sites/matthewherper/2012/02/10/the-truly-staggering-cost-of-inventing-new-drugs/. Accessed 28 Oct 2014.

Harper M. How much does pharmaceutical innovation cost? A look at 100 companies. 2013. http://www.forbes.com/sites/matthewherper/2013/08/11/the-cost-of-inventing-a-new-drug-98-companies-ranked. Accessed 28 Oct 2014.

Hayashi K, et al. Impact of biomarker usage on oncology drug development. J Clin Pharm Ther. 2013;38:62–7.

Hughes B. 2007 FDA drug approvals: a year of flux. Nat Rev Drug Discov. 2008;7:107–9.

Hughes B. 2008 FDA drug approvals. Nat Rev Drug Discov. 2009;8:93–6.

Hughes B. 2009 FDA drug approvals. Nat Rev Drug Discov. 2010;9:89–92.

Hunter J, Stephens S. Is open innovation the way forward for big pharma? Nat Rev Drug Discov. 2010;9:87–8.

IMAP. Global Pharma & Biotech M & A Report 2012. 2012. http://www.imap.com/imap/media/resources/Pharma_Report_2012_FINAL_2F6C8ADA76680.pdf.

Kaitin KI, DiMasi JA. Pharmaceutical Innovation in the 21st Century: New Drug Approvals in the First Decade, 2000–2009. Clin Pharmacol Ther. 2011;89(2):183–8.

Kneller R. Nat Rev Drug Discov. 2010;9:867–82.

Kola I, Landis J. The importance of new companies for drug discovery: origins of a decade of new drugs. Nat Rev Drug Discov. 2004;3:711–6.

KPMG. Future Pharma, five strategies to accelerate the transformation of the pharmaceutical industry by 2020. 2011. http://www.kpmg.com/Global/en/IssuesAndInsights/ArticlesPublications/Documents/future-pharma.pdf. Accessed 28 Oct 2014.

LaMattina JL. The impact of mergers on pharmaceutical R&D. Nat Rev Drug Discov. 2011;10:559–60.

Lessl M, et al. Grants4Targets–an innovative approach to translate ideas from basic research into novel drugs. Drug Discov Today 2011;16:288–92.

Mullard A. 2010 FDA drug approvals. Nat Rev Drug Discov. 2011;10:82–5.

Mullard A. Economists propose a US$30 billion boost to biomedical R&D. Nat Rev Drug Discov. 2012a;11: 735–7.

Mullard A. 2011 FDA drug approvals. Nat Rev Drug Discov. 2012b;11:91–4.

Mullard A. 2012 FDA drug approvals. Nat Rev Drug Discov. 2013;11:87–90.

Mullard A. 2013 FDA drug approvals. Nat Rev Drug Discov. 2014;13:85–91.

Munos B. Lessons from 60 years of pharmaceutical innovation. Nat Rev Drug Discov. 2009;8:959–68.

Munos B. Can open-source drug R&D repower pharmaceutical innovation? Clin Pharmacol Ther. 2010;87:534–6.

Owens J. 2006 drug approvals: finding the niche. Nat Rev Drug Discov. 2007;6:99–101.

Pammolli F, et al. The productivity crisis in pharmaceutical R&D. Nat Rev Drug Discov. 2011;10:428–38.

Paul SM, et al. How to improve R&D productivity: the pharmaceutical industry's grand challenge. Nat Rev Drug Discov. 2010;9:203–14.

PhRMA. Pharmaceutical Industry 2013 Profile. 2013. http://www.phrma.org/sites/default/files/pdf/PhRMA%20Profile%202013.pdf. Accessed 28 Oct 2014.

PWC. From vision to decision Pharma 2020. 2012. www.pwc.com/pharma2020 (2012). Accessed 28 Oct 2014.

Remnant J, et al. Measuring the return from pharmaceutical innovation 2013. http://thomsonreuters.com/business-unit/science/subsector/pdf/uk-manufacturing-measuring-the-return-from-pharmaceutical-innovation-2013.pdf (2013).

Sams-Dodd F. Target-based drug discovery: is something wrong? Drug Discov Today 2005;10:139–47.

Scannell JW, et al. Diagnosing the decline in pharmaceutical R&D efficiency. Nat Rev Drug Discov. 2012;11:191–200.

Schacht WH. Drug patent expirations: Potential effects on pharmaceutical innovation. CRS Report for Congress. http://ipmall.info/hosted_resources/crs/R42399_120302.pdf (2012).

Schuhmacher A, et al. Models for open innovation in the pharmaceutical industry. Drug Discovery Today 2013;18:1133–7.

Swinney DC, Anthony J. Nat Rev Drug Discov. 2011;11:507–19.

Zhong X, Mosley GB. Mission possible: managing innovation in drug discover. Nat Biotechnol. 2007;25:945–6.

Chapter 3
The Importance of Understanding the 'Lived Experience' of Patients in Pharmaceutical Development Programmes

Kay Fisher

Introduction

Improving the 'patient experience' is a hot topic, but capturing the patient's voice early on in the development cycle is also crucial to drug effectiveness. In this chapter, we look at the 'real-life' trade-offs patients are making around their treatment programme and consider ways of using these data to improve the process of drug development. We challenge existing codes of practice, regulatory guidelines—originally designed to protect the patient, but which now seem to be stifling their very importance voice.

Here, we seek to challenge the current thinking on the data inputs that are currently used to guide decisions around drug development. The pharmaceutical world is defined by the condition and disease, and by the scientific impact of drugs, within a controlled environment.

However, our ultimate customers—our patients—live in a world, which is not defined by their condition, but where life takes over, and the impact of their unscientific behaviours around their treatment regimes can literally be the difference between life and death. We are beginning to see a 'taste' for gathering real-world data now, which is a positive step forward, but these new data sets are still clinically led, rather than patient led; the 'patient record' is still completed by clinicians, the treatment reviews are written with a clinical bias. Where is the patient's voice in all of this? Has anyone asked the patients what success looks like for them? Where is it captured? When a clinician meets the patient, there are two experts in the room: the clinician who can guide or lead a patient to an informed decision around therapy choice, or treatment types, and the patient who is the only one who understands how their life choices, behavioural habits, and psychological approach to health might

---

K. Fisher (✉)
Experience Engineers, Buckinghamshire, UK
e-mail: kay@experienceengineers.co.uk

© Springer International Publishing Switzerland 2015
P. A. Morgon (ed.), *Sustainable Development for the Healthcare Industry,*
Perspectives on Sustainable Growth, DOI 10.1007/978-3-319-12526-8_3

play a part in the success or otherwise of this therapy. The focus has always been on ensuring that the clinician is equipped with the best possible evidence to make informed prescription decisions, but there is a case to develop tools and educational materials for patients too, guiding them towards 'best practice' behaviours, and ensuring that they understand the impact of these behaviours on their health outcomes.

How Can We Make Sure We Look Beyond the Science and Capture the 'Human Being' in Our Data?

There is clearly a lot of work going on in the 'social research' space, many pharmaceutical companies are investing in patient engagement programmes which are designed to examine and improve the relationship a patient has with their drugs or treatment regime. Increasingly, the industry is factoring in new data around how we live our lives, and working hard to uncover the truth about those individual lifestyles and behaviours that affect outcomes, alongside the clinical evidence base derived from the clinical effectiveness of the data.

It is crucial to build in these additional layers of data if we want to address the whole truth. As patients, we all make our own decisions about the sacrifices we are prepared to make for the sake of our health—and ultimately that decision does rest with us as individuals. We make trade-offs all the time in our life choices, and our health is no exception. There are thousands of examples of the trade-offs some patients are making around their treatment regime, which, if known, can have a huge impact not only on the narrative that doctors might have with their patients but also on the way drugs are developed.

Exploring the lived experience of patients, I have met a woman who has been told that she has glaucoma and may become blind, but she would not take her eye drops on a regular basis, because they make her eyes go red, and she does not want to be seen looking as though she has been up all night. A more detailed review of her experiential journey reveals that her consultant chooses to soften the blow of the 'brutal truth' at the very beginning of her clinical treatment, in order to reduce the patient's anxiety. Thus, her attitude to treatment is based on her current asymptomatic state, rather than any future state. In this case, 'staying the same' is 'making progress' but this message was missing in earlier consultations.

A teenager who would not take his methotrexate because he knows how nauseous it makes him feel, and he would prefer to be focused for his school day, even though he knows that this might lead to more pain further down the line. He has suffered from this particular chronic condition since he was a young child, and has learnt how each medication in his regime impacts on his physical and mental state. He is willing to trade physical pain for mental clarity because his focus is on the 'here and now' of life, rather than any possibly physical deterioration further down the line. This is how most teenagers deal with all of their decision-making, health is no exception.

A working woman with cystic fibrosis who has chosen to sacrifice regular early morning preventative treatment in order to pursue the career she loves, knowing that she is shortening her life expectancy. This patient has had a lifetime of heavy regime burden, getting up 2 h before school in order to go through a physiotherapy regime, and take all of her required medications. She knows her life expectancy is short, but wishes to live a fulfilled life nevertheless—for her, an additional 2 h of therapy every morning can only be tolerated if there was strong evidence that her life-limiting condition can be cured.

The social science is surely as important as the clinical science here. When we are all patients, the decisions we make are not necessarily logical, but are largely driven by emotion and a need to stay in control of our lives. On occasions, the fact that a drug might save our lives just is not a good enough reason to take it, if it affects the way we want to live.

Is the Pharmaceutical Industry Brave Enough to Let Patients Contribute to Driving the Research Agenda?

If we understand these influences, once a drug is being used in the marketplace, why cannot we gather these data further back in the development process? There are very few social research studies being used to complement the scientific research, yet we are relying on unpredictable human behaviour for the science to work effectively. Where is the patient's voice in the R&D departments? If we gather lifestyle data at the very outset of a drug's development, and throughout its journey to the market, we will all have a better understanding of how the regime, side effects, or delivery mechanism will impact on efficacy in the real world. We will understand how to maximise a patient's engagement with his or her treatment. Ultimately, this critical patient insight may even influence *what* goes into the R&D programmes.

So How Do We Facilitate This?

Firstly, we have to get the timings right. Many clinical trials capture quality-of-life data, but it is either too broad, and uses fairly crude tools which are not individualised to specific disease areas and therapies, or it is captured at the wrong time—we need to understand quality-of-life dynamics *throughout* the treatment experience and not just at the end. This will give us far more granularity and insight around the small but significant decisions patients might make around their therapy regime.

Secondly, we have to find better ways of formalising the capture of this data; in many cases, it is already captured in nurse notes, as it forms an important part of the therapeutic care given to support a condition, as opposed to the clinical instrument which is the subject of the trial. The data need to be 'lined up' alongside the clinical effectiveness data throughout the trial period in order to uncover any 'real-world' behaviours.

Thirdly, we have to develop some end points for these type of data. Clinicians often block these data because they do not know what to do with it, or where to point patients for appropriate information, education, or guidance should 'life' be interfering with 'treatment'.

Finally, we need to build trial methodologies which embed the capture and recording of these data. We need to consider the environment in which the data is captured, the demographic mix of the patient cohort, and the balance of qualitative and quantitative data required to build a robust data set. All of which is feasible during trial stage, just as it is in the real world.

Most—if not all—other industries have the end-user as a starting point to developing their strategy, and if we look outside of the world of pharmaceuticals, we see very different approaches to capturing the voice of the customer. Amazon has a laser-like focus on their customers—their entire offer starts and ends with the customer. They make it easy for customers to get on with their lives by merging their own product and services into their lifestyle. The pharmaceutical industry tends to start with disease need, or a population need, followed quickly by the scientific possibilities. However, it is our idiosyncrasies as human beings, and our life context, which can play a very big part in contributing to the success or otherwise of a certain therapy or drug regime. We see this especially with the 'health-active' population, a customer segment who are the early adopters of preventative medicine and tools to prevent health deterioration. Their attitude to health is purposive rather than responsive, and they are growing in number—a shift that other industries such as mobile, leisure, and sport brands have capitalised on.

How Can We Benefit From This Sort of Research?

There are real benefits of introducing the patient's voice earlier on.

The earlier we can start to uncover more data around personal preferences, the earlier a product is tailored to a patient's needs, and the fewer problems have to be dealt with further down the line. We have to create a better environment where information flows from patient to provider and through to pharmaceutical development from the outset—human patient data, not just clinical data sets.

If we can invest more in this data flow, we can build real 'patient value' throughout the process, and we will know so much more about the things that matter most to patients—things which can have a huge influence on their response to treatment, and potentially offer more choices with different trade-offs, which will make treatment feel more individualised.

Building the Value Chain With Patients Playing a Key Role

The move to build patient-reported outcome tools into the early return-on-investment (ROI) decisions around investment in drug development, or affording reimbursement, is slow and cumbersome. However, with the advent of efficient measurement criteria over long periods of time, using a matrix of data gathered from patients themselves; patient-reported outcome measures (PROMS), alongside a much broader view of health-care costs, we will be able to factor the 'societal cost' of treatment in a more holistic way. Any attempt to create value for the pharma industry, will need to acknowledge the holistic benefits of a treatment regime, alongside the clinical outcomes. It is only when these two are viewed side by side that we will be able to access a true cost benefit analysis.

If the pharma industry lacks the skills to build these tools, then they should consider partnering with those who do, and triangulate these collaborations with those who are delivering the health care. A solution which measures patient impact, clinical efficacy, and payer value is the target here, as it is alignment of all three which will (a) drive maximum engagement and (b) deliver win/win/win benefits. These solutions should seek to individualise as much as possible. Currently, there are no specialist tools measuring PROMS, only generics, which are far too broad in their application, and do nothing to support the shift to individualised therapy solutions.

So What Can We Do to Drive Change?

In a way, the codes and regulations, which are in place to protect the patients, are stifling their voice. Approaching patients for their views has always been very difficult for the pharmaceutical industry, and they have therefore used the clinician's voice as a proxy for what patients value most. However patients spend very little time with their clinician, and most of their time living their lives outside of the clinical environment, so it is the patients who must be cast as the experts here, and it is the 'lived experience' data, which we must seek to capture. Clinicians will benefit from this too—whenever they are exposed to these data, they value it, address it, and act on it. In our earlier example of a patient with glaucoma, when the consultant was confronted with the patient's need for a 'brutal truth' consultation at the outset, plus a more flexible approach to regime to avoid red eyes during the working day, he changed his narrative during initial consultations. When a nurse is negotiating the tricky balance of encouraging a teenager to become more compliant generally, while retaining their engagement in long-term treatment, she changed her consultancy narrative to enable more formal 'benchmarking' of the teenager's knowledge and understanding of each of his therapies. This instantly led to improved engagement, and was the beginning of a more informed dialogue between the nurse and patient going forward. The pharmaceutical industry can help with this; they can build this softer, yet essential life science insight into their patient data. If we really start to understand what success looks like to the patient, as compared to the scientific outcome, we might be very surprised, and we certainly should not presume to know without first consulting them.

The new world has to engage patients as an equal third-party stakeholder, whose interests may sometimes be aligned with health providers and sometimes aligned with pharma, or both, or neither. Ultimately, with this framework in place, we might make better decisions around what drugs are developed and how that development evolves. A good place to begin this change would be with rare diseases—it would not be difficult to talk to every single patient in these cases, thus ensuring 100% sample rate.

Why Bother?

From a purely commercial point of view, if patients influence and support drug development from the start, then it ticks regulator and purchaser boxes ahead of time, and makes it harder to resist at later stages. Industry's leading best practice would always feature the patient's voice throughout drug development. This is not exploiting patients, who can and will decide for themselves how vocal they want to be. One of the biggest issues is persuading patients that their lifestyle issues really matter, that drugs cannot really be developed in a scientific vacuum if they are to be optimally effective, so we really do need their input.

Patients as equal partners in the development process? It seems ridiculous does it not, what do patients know about science? Yet, they often have the casting vote in the success or otherwise of the drugs they take, because of the choices and trade-offs they make on a day-to-day basis. The sooner we all understand the value of this, the sooner we can evolve the pharmaceutical industry model into one that puts patients at the heart of its work.

Chapter 4
Listening to the Voice of the Patient to Facilitate Earlier Access to Promising Medicines: Interview with Sjaak Vink

Pierre A. Morgon

The discussion with Sjaak Vink provides a fascinating account of what can be accomplished when the industry gears itself properly to work more closely with the patients or the nongovernmental organization (NGO) representing them. The insights are particularly relevant not only to the management of clinical development but also to the global corporate governance as they address the need to evolve management practices, towards a long-term strategic orientation, openness to real-world data, and responsiveness to societal pressure. The industry will need to flex its procedures, to open up to broader collaborative approaches, and to foster a company-wide orientation towards innovation.

Pierre A. Morgon: What's your opinion about the role and place of sustainable development in the health-care industry? How does it contribute to its evolution and what it means in your scope of accountability?

Sjaak Vink: Echoing the concepts illustrated by Jim Collins, in turbulent times more than in stable times, a greater attention should be paid to timeless fundamentals so that companies are built to last. These fundamentals include the recognition by companies that they have great responsibilities, including the preservation of freedom of entrepreneurship and freedom of mind, and the contribution to a sustainable society.New concepts such as transparency, sharing of knowledge, and open innovation are increasingly taking root in the corporate world, even in the largest companies. But beyond the understanding, companies have to act upon it....More spe-

---

P. A. Morgon (✉)
Theradiag, Croissy-Beaubourg, France
e-mail: pierre.morgon@wanadoo.fr; mrgn@bluewin.ch

Eurocine Vaccines, Solna, Sweden

AJ Biologics, Kuala Lumpur, Malaysia

cifically looking at the pharmaceutical industry, the main obstacle to overcome is the tremendous amount of distrust that this industry has generated in the lay public. This is both surprising and discomforting, as when one works closely with industry staff involved in the development of innovative medicine, one meets intelligent and committed people who put their minds and their hearts in the search for new cures. The mind-set changes when you look at the upper echelons of the management, as the top layer is concerned by short-term shareholder value. I'd posit that the senior management of these companies would be better off if it was setting free the intellectual spirit of their staff. In other words, the health-care industry needs shareholders that take a different view at value creation and in order to get there, it needs brave leadership that makes them understand the value of thinking differently, of taking a longer-term perspective and of creating a unique selling proposition resting on value for society. Critics of such longer-term orientation often argue about cash and profitability. Arguably, money is the fuel and as such, it remains important, but the most important is the momentum, the "why" that has to be answered with deeds and not just with words. I understand that it can be difficult to obtain approval internally for long-term focused initiatives, as the costs are borne immediately while the return is often hard to assess and is expected to materialize in a longer timeframe. But beyond the classical "business case," there are other forces at play that will increasingly drive decisions, and in this respect one shouldn't underestimate patient power, amplified by social media. There have been several recent examples of this rise of patient power. Think for instance about the melanoma patients who, supported by their treating physician, challenged the fact that they have been denied enrollment in clinical trials on melanoma. They took their cause to Twitter and Facebook and they attracted so much attention that one of them was invited to the David Letterman Show; the discussion shamed the company involved in the trial and the magnitude of the awareness triggered moved America. Another force that increasingly drives the decision and pushes shorter timelines is the nowadays treating physician. Being enabled to be up-to-date on latest innovation 24/7, this caring physician demands for the best possible treatments as soon as possible. In the US, healthcare professionals were questioned on the timelines for new medicines. Over 75% of oncologists thought it takes too long before they are allowed to work with latest developments. Furthermore, they argued Real-World Data collected by them while treating their patients might appear to give valuable insights for pharma companies, payers, and regulators on treatment patterns, patient characteristics, quality of life of patients, quality of life of patients, et cetera. And on outcome and duration of treatment required. As PatientsLikeMe data already proves us. New grounds to explore for the better of all. In the future, we will witness more of such pressure to ensure earlier access to medicine if the complex and time-consuming approval procedures for new medicines don't change. The pressure will also be exerted through other conduits. In addition to the patients' voices being heard on the social media, you can think about the pressure applied on company employees in their lives as citizens, in their family circle, but also when with their friends or in any other social interaction context. Recently, I visited the Innovation for Health Congress in Barcelona where I spoke with Nigel, a scientist from one of the big pharma companies. He told me that the younger employees within his company see potential for innovation. I think this generation could force a change of the system. I'm convinced this "Generation Next" will reach out for a tipping point. Within the companies they work, within the industry. Young researchers, scientists, and physicians creating a new moral awareness within which it goes without saying that patients come first. As it was meant to be in the earlier days by the founders of the Merck's and Johnson & Johnson's of this world. The industry needs to flex its procedural attitude. The staff within the companies is increasingly challenging their senior management. We at myTomorrows are striving to engage them and to join forces so that we reach a tipping point.

PM: **Could you describe sustainable development initiatives you've been working on with pharma companies? What were the hurdles to overcome and practices to apply? How eventually does it bring value? What do you think would be important to implement them successfully?**

S Vink: The first observation is that all companies are not alike in this respect; a short and simplistic answer would be that small and innovative companies are easy to work with, while large ones are not since they are driven by accountants, bankers, and lawyers. But the reality is more complex than that, and a culture of innovation plays a pivotal role: Companies that have such a culture tend to behave and interact in an entrepreneurial way, irrespective of their size. Overall, myTomorrows has good relationships with the small companies. But at times, you see small, innovative companies which add board members coming from the "big pharma" world and you witness some change as these big pharma former executives introduce a lot of the old, stiff reasoning. And as a consequence, such small companies are not easy to partner with. Luckily, most of the time this is not the case. Also, there are some large companies that are genuinely trying to work with us and we are managing to run projects together (for instance, we are working with a company on a project for a treatment for cancer). These companies become more and more interested to work fluidly with a patient-centric organization like ours. In all instances, myTomorrows is willing to retain its independence and to ensure that patients and doctors get the transparent, reliable, and trustworthy insight that they need. Partner companies will not be allowed to jeopardize our speed of execution. The question that we're indirectly asking to our partners is "are you ready to move fast?", and the corollary—yet unspoken—question is "can you move fast?"

PM: **What would you recommend the industry should change towards ensuring sustainability?**

S Vink: Within the industry habitat, there are other stakeholders and a number of them are also displaying clear conservative traits, such as some "key opinion leaders" and sometimes the medical associations. I challenge them to be part of the growing movement that doesn't accept the status quo and aims to make innovative medicines more accessible for patients with unmet needs. For instance, in the field of oncology, the ASCO (Note: American Association of Clinical Oncology) has recognized that things have to change in terms of the way the patient voice is factored into drug development and access. This is really encouraging. It's important that the association dares to lose control and organizes the evolution in such a way that the voice of society is leading. If this is what they have in mind, the association will realize their aim to serve society. "Are we daring enough to be humble," ASCO President Clifford Hudis asked an audience of oncologists, "and serve our patients?"I would recommend to the industry to tone down its obsession for control. Unquestionably, safety is important, but let us not allow it to go too far. At the moment, the authorization of innovative treatments takes 15 years. Seriously ill patients don't have this time. My goal is to contribute to a growing number of patients that get access to promising medicines. When one looks at companies such as Google, Facebook, eBay, etc., they create an environment in which it is free to go and to make use of it. From the user's vantage point, it is a noncontrolling environment which it's upto the user to act. In the medical world, a decision to act should be taken by the physician and his patient. Without delving in the motives of these companies, one has to recognize that this is very positively perceived by the lay public, while the willingness to retain control and the paternalistic tone of the pharma industry is not going down well with society. Industry bosses should act less as managers and more as enabling leaders, freeing up the willingness to innovate and to go the extra mile in their teams with passion.

Chapter 5
Drivers of the Real-World Data Revolution and the Transition to Adaptive Licensing: Interview with Dr. Richard Barker

CASMI (Center for the Advancement of Sustainable Medical Innovation)

Pierre A. Morgon

The discussion with *Richard Barker* addressed the concept of adaptive licensing and the consequences for the industry in terms of implementing a genuinely patient-centric strategy. It is striking that the industry appears as more conservative than the regulatory authorities when it comes to exploring new licensing routes and that a change of mind-set is needed as much as an evolution of the industry's organization. The required evolution encompasses a greater integration of industry functions along the product life cycle, the focus on better profiled patient populations, taking into consideration both the clinical and behavioral dimensions, and a commitment to deliver the expected outcomes.

Pierre A. Morgon:	**What's your opinion about the role and place of sustainable development in the (thematic) in the health-care industry? How will it evolve in the next few years?**
Dr. Richard Barker:	The historical process of drug development is not sustainable. When looking at the building blocks of the value chain of the industry, a new ecosystem is already taking shape at the discovery level. It is resting on more sharing of data and information and is more competitive in the way the partnerships with the academic world are designed and executed. Therefore, the discovery is already remaking itself.
	When considering the clinical development, one can see that "the ice is cracking." The previous model, based on rigidly

---

P. A. Morgon (✉)
Theradiag, Croissy-Beaubourg, France
e-mail: pierre.morgon@wanadoo.fr; mrgn@bluewin.ch

Eurocine Vaccines, Solna, Sweden

AJ Biologics, Kuala Lumpur, Malaysia

defined phases, could still be applicable to some therapeutic segments but is definitely too rigid. A more adaptive world is in the making, as illustrated by the more flexible regulatory dialogue made possible by the FDA's four exceptional routes (fast track, breakthrough therapy, accelerated approval, priority review) and the EMA's three exceptional ones (conditional approval, exceptional circumstances, accelerated assessment). For the time being, these routes have been used primarily for oncology and rare disease drug candidates but over time, they will have an impact on other diseases.

For instance, one can observe that the clinical development of PCSK9 agents in dyslipidemia is resting on 10+ phase III trials addressing well-defined patient populations, hence very different from the larger-scale clinical studies performed for statins, and illustrating a different clinical and regulatory pathway to address the full potential of the drug candidate.

Overall, the regulatory process is increasingly flexible and further ahead compared to the industry's more conservative attitude. Some health-care systems are already taking an innovative and collaborative approach to discussing the required changes; one such example is the Health Innovation Network in the UK, within which the NHS is working with the industry. The approach was initiated by the government and exemplifies a change in the mind-set, which could be emulated in other countries.

Regarding the pricing and reimbursement phase, there's a stepwise integration of evidence of value in clinical development, yet this is taking longer than it should since the two key industry functions involved, clinical development and market access, are often completely isolated in their respective silos and report into different echelons of the senior management (most often R&D and Marketing, respectively). As a consequence, there is often a major disconnect between the clinical development program and the collection of value data. This is compounded by the fact that such data vary in nature and relevance across countries, and it is not possible to satisfy the payers across the various health-care systems with a single, unified clinical development program.

As a consequence, value determination and value realization becomes the new bottleneck in the path to commercialization of novel therapies. One solution is to integrate value in the design of the clinical development, with an early crafting phase involving the input of regulators, payers, and patient associations, so as to ensure the collection of both clinical and value data.

PM: Could you describe succinctly the sustainable development initiatives in the health-care industry you've been involved into? What were the specific hurdles to overcome and practices to apply? How eventually does it bring value? What do you think would be important to implement successfully?

Dr. Barker: A good example is the project "Get Real," which is about the use of real-world data in clinical development to help establish the value of the product, by determining effectiveness in real-life conditions.

The challenge for the industry is the "one-time" fixing of the price and the deeply rooted belief that prices and reimbursement levels can only go down. It would be important to consider a system in which prices could go up too, which would be a real revolution. Until the payers show some genuine willingness to revisit the prices and increase them in light of compelling real-world data, the industry will be driven to set the price as high as possible at the beginning of the process so as to cushion the risk of price erosion.

This has to be put in perspective with the growing number of presentations in conferences and papers that converge to recommend that the industry changes its marketing and sales model and shifts its emphasis on the relationships with the payers, as well as between the industry and the patients. Patient power and patient transparency are increasingly critical. The industry shouldn't be obsessed by promoting to the patient, but should rather focus on counseling and on the procurement of services truly valued by the patients, within the boundaries of what is permissible.

It raises the perennial question of the lack of trust, which has been visible for many years and the belief persists that the industry cannot be trusted. As a consequence, it is harder for the industry to deal with patients, it required great care and sense of responsibility, and it could be better managed through a trusted third party.

From the perspective of the industry's organization and processes, integrating clinical research and value requires multidisciplinary international teams, fully accountable for the success of registration, pricing, and reimbursement, and such teams should remain in place throughout the entire life cycle of the product. The industry should be reconsidering its approach to clinical development by segmenting patient populations according to clinical response on a continuous basis, as the means to profile the patients and segment them become available. That would create a platform on which the industry could gain its revenues according to outcomes rather than volume by allowing products to be prescribed to and used by the right patient segment. Also, the industry should revisit completely its stakeholder's management approach and focus on key decision makers and address them specifically as customers (e.g., NICE in the UK).

When looking at the patient dimension, the industry needs to focus on the behavioral factors which drive adherence, either as accelerators or decelerators. The goal should be to collect evidence as to how to boost adherence, and the exercise is complex since there are many types of patients with different behavioral contexts and psychological barriers.

For the industry, the adjustment is a painful journey because the previous model has been extremely successful. Yet, in the current context of massive financial squeeze of the health-care systems, the gaps between resources needs and availability are such that changes must be drastic. As the industry has demonstrated its prowess at project management, studies tracking, and data mining, it should use its energy, capabilities, and financial resources and act as a willing and powerful partner to solve these problems and implement the relevant changes.

Chapter 6
Sustainable Development for the Healthcare Industry: Vantage Point from Emerging Economies

Satish Chundru

> *The health of nations is more important than the wealth of nations*
> William James Durant.

Introduction

Product quality is only one aspect of the bigger picture that includes product safety as an equally or more important element when it comes to the sustainability of pharmaceuticals. To ensure the safety and quality of health-care-related products, in addition to ethical commercialization practices, has been the never-ending task of various regulatory bodies around the world.

For example, the Taiwan Food and Drug Administration (FDA) that emulates the model of the US FDA, was in the process of implementing a new regulation for all raw material imports of drugs. The suppliers/sellers had to register their product via local Taiwanese partners with the Taiwan FDA by submitting documentation of their product profile. I had first-hand exposure to this process, as I was co-ordinating efforts for the company I was working for, to file for Taiwanese drug master files (DMFs) for future active pharmaceutical ingredient (API) supplies. Later, there was another expedited regulation to provide the product documentation apostilled by the Taipei Economic and Cultural Centre to issue import permits. (These announcements were made in Chinese on the Taiwan FDA website (Taiwan FDA 2013) with no other international references.) This is a welcome change as it would ensure compliance and thus, the goal of safety along with quality is achieved. Regulations have become highly essential and without these, it is hard to imagine how the health-care industry would have evolved in the global market place. Besides, regulations in several cases are also legally binding, which forces the companies to adhere and maintain a high level of transparency and stability.

S. Chundru (✉)
Northeastern University, Boston, USA
e-mail: chundru.satish@gmail.com

© Springer International Publishing Switzerland 2015
P. A. Morgon (ed.), *Sustainable Development for the Healthcare Industry,*
Perspectives on Sustainable Growth, DOI 10.1007/978-3-319-12526-8_6

> When we look at product pricing, we observe in emerging markets, least developed countries (LDCs) and low-income countries (LICs), the price of multinational corporation (MNC) products is on the higher side compared to the prices of products from Emerging Market Players (EMPs). Many, if not all, of Sandoz's and Mylan's generics are priced higher than those generics from local manufacturers/home-grown companies.

In this chapter, we are going to explore how sustainability in commercializing pharmaceuticals could be achieved by focusing on regulations and effective governance.

Sustainable Commercialization

Sustainable commercialization is the process of commercializing products in a way that does not disturb or destroy the biosphere and the econosphere we habitate. What this means is commercialization, which encompasses several processes like production of raw materials, packaging, distribution, and selling should be done in a thoughtful and meaningful way that just does not operate on the objective of revenues and profits, rather a core purpose, which is the long-term value creation for all stakeholders instead of only the shareholders should be the goal.

> Discussing how investors behave in emerging economies is interesting, especially profit booking being prevalent in emerging markets, after the stock hits its new high or does better than the past few trading sessions, many investors off load their positions, unless there is a clear bull run or a new product approval. Though this behaviour is not exclusive to pharma and health-care stocks, this has been the case across.

Roles Played by MNCs Versus Local Manufacturers in Emerging Economies

Emerging markets have become the hot favourites of MNCs for the past few years, with Bayer and Sanofi leading in 2013 in emerging markets revenues (Top 10 Drugmakers in Emerging Markets 2013). MNCs and home-grown/local companies play varied roles on the ground with each contributing and addressing a different need within the broader need of supplying quality and safe medicines.

> When we look at product pricing, we observe in emerging markets, LDCs and LICs, the price of MNC products is on the higher side compared to the prices of products from EMPs. Many, if not all, of Sandoz's and Mylan's generics are priced higher than those generics from local manufacturer's/home-grown companies.

The below table points out that different players have different cards to play (MNCs versus EMPs) in the healthcare business. Local companies with their strong presence and adjustability to the existing business environment play a major role or a more important role than MNCs in meeting requirements of patients that form the major part of mid-low-income countries. MNCs on the other hand have got resources and innovative medicines that can tend to niche segments of patients.

Contrasting MNCs and EMPs in emerging economies		
Element[a]	MNCs	EMPs
Presence	Weak	Strong
Global network	Strong	Weak
IP (intellectual property)	Strong	Weak
Compliance to regulations	Strong	Weak
Adjustability[b]	Weak	Strong
New products	Strong	Weak
Pricing	Weak	Strong

[a] The weak and strong status of MNCs and EMPs in the above table are relative to one another.
[b] Adjustability refers to be able to adjust business to a great degree to operate in the local environment.
MNCs multinational corporations, *EMPs* Emerging Market Players

We have seen quite a deal of consolidation of businesses between these two groups in the last decade, for example: Abott's Piramal acquisition, Sanofi's Shantha acquisition, Takeda's Multilab and a few others. There have also been acquisitions of companies based in advanced economies by EMPs (global players from emerging markets). This could be attributed to leveraging each other's strongholds, for example, when an EMP acquires a company in an advanced economy, the EMP gets access to the market that supports the better price for products; likewise, when an MNC acquires an EMP, it gets access to a low-cost manufacturer, new market, existing setup, human resources that are aware and have been operating in local business environment.

> MNCs have faced their share of problems by acquiring EMPs. Integration is difficult post acquisitions due to no match between existing systems or processes between these two groups besides cross-cultural issues. There have been incidences in product quality failures, Sanofi's Shan5 (a product of Shantha Biotechnics acquired by Sanofi) was disqualified by the World Health Organization (WHO) following quality complaints or Ranbaxy which serves as a classic example for how things can go wrong, when Daiichi Sankyo acquired Ranbaxy, irrespective of what their due diligence revealed, they were ready to take a huge risk and enter unfamiliar waters, they went forward with the acquisition and years later, they ended up off-loading the company to another Indian company. Daiichi Sankyo took a big financial loss on this transaction besides loss of reputation and brand value.

The elements mentioned in the above table are dynamic and changing often with actions being taken by companies as a part of their strategy and the need to survive and deliver values. This is also dependant on overall global business and economic sentiment and changes occurring in the policy or guiding frameworks of business operations in various countries.

Presence EMPs have a solid presence in their countries of origin whereas MNCs are still to get their presence to match that of EMPs. Most of the EMPs begin their business manufacturing products that are less technically intense and graduate to advanced products as they gain capabilities in the form of capital and human resources.

Global Network MNCs have got an established presence in many countries around the world. EMPs' primary turf is home and once they reach size and means, they expand to foreign countries at least in the form of distribution points. We can find many EMPs that do business globally, for example, Dr. Reddys, Sun, and Abdi Ibrahim.

Intellectual Property MNCs thrive on strong intellectual property (IP) whereas EMPs primarily work with making copies of the innovator's products with an aim to launch post loss of exclusivity or patents and compete on price to gain market. Thus, EMPs have their stronghold on secondary IP in process patents, polymorphs and salts.

Compliance MNCs have global teams in place to check their compliance-related activities, as they are under constant scrutiny from various sources. Though EMPs comply to regulations, their focus is rather on growing revenues than on adding value.

Adjustability EMPs with their knowledge on local governance adapt their business to suit the existing environment compared to MNCs which struggle to align policies and processes. When EMPs expand their business beyond borders, they too find it difficult to adjust their business.

New Products This is home ground for MNCs which bring out innovative medicines often those addressing unmet medical conditions or rare diseases. In contrast, EMPs have not got the resources to indulge in research at this level to bring out innovative medicines in terms of both investment capital and scientific know-how. They rely mostly on generic versions of the innovator's products.

Pricing The costs of EMPs are lower than those of MNCs. The key contributor to costs besides material costs are those of wages and salaries that are on the lower side for EMPs. To offset this and price their products competitively, MNCs build capacities in emerging economies that sell locally.

Vantage Point

The discussion for this chapter is based on taking a vantage point into the emerging economies, which are fast changing, and how they respond to sustainable commercialization, if any. The factors affecting sustainable development and in turn sustainable commercialization have been narrowed down to regulations and governance, these being the drivers that operate at a bigger picture level towards the larger good. Government regulations and legislative requirements are significant drivers as any changes enacted in this direction are tracked by companies and this paves the way for their increased awareness which eventually going forward leads towards sustainable development for the industry. This view is supported by the research conducted by Francois Leeuw and Ilse Scheerlinck of Vrije University in Brussels who studied the factors that shape companies' corporate sustainability.

Regulations

More often than not, the regulations are unclear in emerging economies. Governments and regulatory bodies often initiate something which is not well planned and structured; the outcomes of such initiations are limited or nothing. This puts companies that already struggle with aligning their global regulatory strategy across markets in a tough spot. However, changes are happening at varied speeds, for example, the move towards electronic common technical document (eCTD) format for submissions is growing across the countries and this makes one regulatory aspect easier.

The seven regulatory barriers that need to be overcome in emerging economies are western approval, local clinical development (LCD), certificate of pharmaceutical product (CPP), good manufacturing practice (GMP), pricing approval, document authentication and harmonisation (Wileman and Mishra 2010). Regulatory compliance along with harmonisation stands out as the barriers that are more relevant in this context of sustainability.

Harmonisation

Harmonisation of regulations should encompass and extend to all aspects of operations that are undertaken to bring a product to market and continue supplying it to different groups of customers; this includes, but is not limited to, regulations of clinical trials, pharmacopoeias, laboratory practices, manufacturing processes, site inspections, delivery/distribution practices and industrial safety practices.

Harmonising these regulations to the greatest possible extent would be a rewarding task for all stakeholders. This would pave the way to eliminate duplication of efforts especially in clinical trials, studies and let the best use of resources, which are almost, always limited be it time, capital or well-qualified human resources. Other tangible and intangible results of harmonisation are reduced development times, less cumbersome approval processes across countries, decreased time to market which means faster availability of key medicines to patients (Honig 2013).

> Regulatory inspections followed by reinspections of manufacturing and clinical trial sites could lead to unnecessary or duplicated efforts and consume critical resources. A multinational pharmaceutical company and those companies selling to companies in foreign countries (business to business, B2B) will go through multiple inspections from different countries and from different regulatory authorities in different regions. Food for thought of this practice is if there is any incremental value of the repeated inspections aimed with the same objective. If these duplicative tasks and efforts towards organizing and facilitating multiple audits and inspections could be eliminated, resources could be diverted to other critical aspects like safety, quality and efficacy.

There has been significant progress in the area of harmonisation since such initiatives were started by various organizations like International Conference on Harmonisation of Technical Requirements for Registration of Pharmaceuticals for Human Use (ICH), Pharmaceutical Inspection Convention and Pharmaceutical Inspection Co-operation Scheme (PICS), African Medicines Regulatory Harmonization (AMRH), Pan American Network for Drug Regulatory Harmonization (PANDRH), Organisation for Economic Cooperation and Development's Mutual Acceptance of Data (OECD's MAD), International Medical Device Regulators Forum (IMDRF). However, the destination is still a long way away and efforts have to be carried out ahead in all aspects especially by both public departments and private organizations that are required in allocating resources for this endeavour.

> ICH has contributed greatly to the world of regulatory harmonisation. In addition to the founding member countries of the USA, Japan and those in EU, ICH has invited permanent members from Regional Harmonisation Initiatives (RHIs; like Asia-Pacific Economic Cooperation (APEC), Association of Southeast Asian Nations (ASEAN), Gulf Cooperation Council (GCC), PANDRH, East African Community (EAC)) to participate in meetings to harmonise regulations across non-ICH countries (ICH 2014).

Compliance

We have discussed how harmonising of several aspects of regulations is worthwhile for the industry and patients. Compliance to regulations or even better, harmonised regulations is vital as the healthcare industry including pharmaceuticals is under increased critical examination from not only regulatory bodies but also B2B customers in many countries. These could be in the form of audits, visits, reviews or investigations. It does not look like it will get any easier; rather it looks like this is going to get more intense with the fact that there still are many occurrences of failed audits and major observations resulting in export bans or calls for other correctional procedures like product recalls. Besides, with the growing social security implementations by several countries, growth is anticipated in public procurement schemes, which calls for rigorous regulatory enforcement and regulatory compliance.

Financial regulatory compliance for those companies that are publicly listed is another vital aspect of companies for business continuity in the long term. Most of the companies are exposed to accounting risks and face a threat that could harm shareholder value in markets. These have led to the rise of chief compliance officers in MNCs and this concept is still at a nascent stage in companies from emerging economies.

Pharmaceutical companies have to adhere to regulations set by bodies at different levels, i.e. regional, national and international and at all stages of the product life cycle right from research and development (R&D), manufacturing operations to distribution, marketing and pharmacovigilance. Ensuring these at all levels in multiple territories is a complex yet essential task for companies.

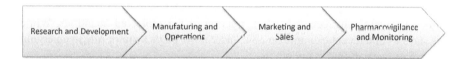

A consolidated value chain in pharmaceuticals that needs to comply to regulations at every stage.

For most pharmaceutical companies in the world with no exception for those arising out of emerging economies, the primarily focus of their business is the USA, the reason for this being, the USA is the world's largest market for pharmaceuticals in addition to being one of the few countries that supports free-market-based pricing. For these companies wanting to do business in the USA, strict compliance is essential now more than ever as the consumer protection branch of the Department of Justice (DoJ) has indicated that current good manufacturing practice (cGMP) enforcement would be a top priority (Burgess and Kang 2013).

> In 2013, Ranbaxy pleaded guilty and agreed to pay US$ 500 million (US$ 350 million to settle False Claims Act (FCA) lawsuit and US$ 150 million for related felony charges), for allegedly selling adulterated drugs that violated cGMPs. This is one of the only two FCA settlements based for failing to comply with cGMPs under the Food, Drug, and Cosmetic (FDC) Act (Burgess and Kang 2013). This case stands as the recent evidence to what could go wrong in cases of non-compliance.

Governance

Good governance is an essential factor for economic growth and sustainable development at all levels and within all sectors of society (Anello 2008). Governance in emerging economies is often overwhelmed by the numerous challenges they face and the limited resources at disposal, often resulting in unnecessary bureaucracy and unclear guidelines and regulations for conducting business. For example, the Indonesian Government's plan to expand social security coverage to the entire 250 million odd population has been unclear from the beginning. The local pharmaceutical manufacturers expected that the government would invite bids for tenders to supply drugs to the government that would be used as a part of this social security scheme. The local companies started discussions and negotiations with suppliers of APIs for the lowest prices guessing that the quantities would be huge. The suppliers started working out detailed plans based on the volume-based business rather than staying on or focussing on the value-based business. However, this ended with nothing ever realizing concretely and the few products that did receive invitations from the government received only a fraction of the projected quantities. While this stays as one recent example on how uncertainty plays a major role in emerging economies, we can find many others.

On the other side of things, there were instances in countries like Mexico where tenders have been issued for public procurement of medicines by government agencies. However, these have faced their set of issues in the way of anticompetitive practices like bid riggings that caused huge losses to government coffers indirectly having an effect on the consumers. In 2010, the Federal Competition Commission of Mexico (CFC) inquired into the public procurement tenders by Mexican Institute of Social Security (IMSS) between 2003 and 2006 and this investigation revealed that several firms that participated in the tenders were involved in anticompetitive practices that resulted in higher procurement prices for the IMSS. Six pharmaceutical companies were penalised for a total of 151.7 million Mexican pesos, which was the maximum amount allowed by law at that time (Competition Policy and Public Procurement 2012).

Under the umbrella of governance, three elements that affect sustainable commercialization to a great extent are compulsory licensing, prevalence of counterfeit drugs and pricing controls.

Compulsory Licensing

In the recent past, we have seen instances where governments in emerging countries have employed various ways to get patented products to their patients. Brazil, India, Indonesia and Thailand, all have at various instances issued compulsory licenses (CLs). Though Trade-Related Aspects of Intellectual Property Rights (TRIPS) and World Trade Organization (WTO) permit this under circumstances of health calamities and disease breakouts, CLs have been used to invalidate IP in regular situations. Governments justified their actions portraying this as making products affordable or to make them accessible and thus improve public health. Companies on their behalf have followed tiered pricing to make way for affordability.

It is interesting to note that most of today's advanced economies including the USA, Canada, France, Belgium have used CLs in the past for different objectives to different extents. Compulsory licensing of medicines in instances of non-urgency has to be limited by governments to safeguard IP and to promote scientific research. The real risk that compulsory licensing could discourage foreign investments and development of advanced technologies by making other economic environments more attractive to firms in technology-exporting countries has to be considered in great detail before issuing licenses. The absence of national and regional systems of innovation and reliance on compulsory licensing can mask structural problems in these countries and make them harder to solve in the long run (Reichman and Hasenzahl 2003).

Another aspect of CLs tends to be the quality of medicines produced by these licensees in emerging economies. The quality of production of these local firms is lower than that of the patent holder. An example of this is the compulsory license of Kaletra issued by the Thailand government to the Government Pharmaceutical Organization (GPO). The quality of GPO's product was subpar and the global fund which granted US$ 133 million to GPO in 2003 to upgrade facilities withdrew support in 2006 after GPO failing to meet WHO quality standards (Bond and Saggi 2012).

One solution to the problem of access and affordable medicines is pooled procurement strategy; this is explained as several emerging economies coming together and leveraging economies of scale in their negotiations with companies for deep discounted prices for the increased market size. This collaborative approach could work in favour of both parties as governments can benefit when these companies make local investments to serve these patient bases and thus also act as means to promote research instead of discouraging it while innovator companies benefit by holding their exclusive rights (Reichman 2009) and to make the pie even bigger could negotiate tax breaks on their investments.

Counterfeit and Spurious Drugs

Counterfeit and spurious drugs besides being a menace for public health also cause enormous losses to companies. This is in addition to the health risks they carry to those who consume. Counterfeit and spurious drugs have been in existence since decades in both advanced and emerging countries. It is estimated that about 15% of the medicines sold globally outside of advanced economies are counterfeit and this rises to 50% in certain parts of Africa and Asia (Sridhar and Gostin 2010). The efforts of various organizations and agencies including International Medical Products Anti-Counterfeiting Taskforce (IMPACT), a WHO global initiative with member countries, to curb these have yielded mediocre results especially because of the failure of co-ordinated governance in this direction.

> There are many statistics on the size of counterfeit medicines or how much in revenue loss they account for and the deaths they cause globally. The important thing next to avoidable loss of life is counterfeit drugs destroying the value of an industry and how technology can be used to fight this by not relying completely on governance or governments to tackle the problem. Sproxil, Pharmasecure and others serve as examples of how existing technology coupled with industrial partnerships could be leveraged to safeguard the interests of drug consumers. The product package is labelled with a scratch-off unique number which could be texted to a hotline and a return message will confirm the authenticity of the product (Rosenberg 2014).

Stringent regulatory control of drugs as well as stringent enforcement by national regulatory authorities contributes significantly to the prevention and detection of counterfeit and spurious drugs (WHO 2012). A good question to re-ask by the governments in emerging economies would be: (1) if the prevalence of counterfeit and spurious drugs is attributed to their cheaper price or (2) inadequate awareness of safety and dangers of consumption of these. Also, a more focussed global collaboration, coordination and cooperation between countries would address the issue of these counterfeit and spurious drugs to a better extent.

Pricing Controls

Price controls of pharmaceuticals are prevalent in both advanced and emerging economies. A major concern of price controls has been its effect on pharmaceutical research, and how it could potentially impact the outcomes of future research. This has been validated to various extents by several researchers, one of whom says that product price cuts of 40–50% would result in reduced R&D projects by 30–60% (Abott and John 2005). In another report by the US department of commerce, the pricing controls employed by the OECD countries resulted in revenue losses estimated at US$ 18–27 billion to pharmaceutical firms which in turn resulted in R&D reduction by US$ 5–8 billion annually. These reductions in billions of dollars in R&D have resulted in an estimated 3–4 fewer new molecular entities annually, which is a great loss to all stakeholders (Pharmaceutical Price Controls in OECD Countries Implications for U.S. Consumers, Pricing, Research and Development, and Innovation 2004). Based on the estimated cost of developing new drugs, trend and rate of approval of new drugs, this capital investment could have paved the way for these companies to come out with a greater number of newer medicines.

Additionally, it has also been noted that drugs account for 12% of overall health-care costs (Farrell et al. 2008), for 7% which makes the cost contribution to the overall health-care expenditure by hospital outpatient and inpatient care, physician and clinical administration and other delivery services much higher than the cost contribution of drugs. Thus, drugs account for a small percentage of overall health-care costs (Hooper 2008). While the above-mentioned figure is with reference to the USA, the case in emerging economies could be comparable. Governments in emerging economies could focus on working towards alternate means to reduce health-care costs, for example universal health-care schemes and mandating health insurance by creating health funds could be explored and economic models could be worked out. This would keep an industry that is so essential alive and growing in addition to promoting competitive research.

The widely accepted DiMasi study puts the cost of producing a drug and releasing it to the market to be around US$ 800 million, which is unlike products in many other industries. This huge upfront cost along with long lead times often with no guaranteed success puts enormous pressure on the companies comprising pharmaceutical industry. With no financial incentive, companies would be reluctant to invest in research, the direct outcome of which could be reduced number of new products. This could turn out as a blow to the many existing unmet medical conditions.

Extensive interference from government by controlling the product prices would result in decreased returns for these companies, most of which are publicly funded and could leave investors wary besides reduced revenues and incomes limiting the overall capital available to carry out critical operations and future research

Closing Comments

Is sustainability/sustainable development really the issue of only the advanced markets with minor or no impact to the emerging economies? A little awakening of the senses, a little observation into the evolution of the industry and its progression into the future, a little understanding of how businesses are no more local and are inter-linked including the elements discussed in this chapter would tell us the answer is a no and so should the efforts addressing them be inter-linked and global (though they start local). Sustainable commercialization and sustainable development in the industry that is at the very front in saving lives are pressing issues globally that would move us collectively into the prospective or the not-so-prospective future.

The environment in which the health-care industry operates is becoming tougher and more stringent not just to safeguard public health but also to keep the actions of companies within the boundaries of compliance. Companies on their behalf would want to come out with solid scientific data/evidence in support of the products they develop which would let them go to market smoothly and as a result provide value to all stakeholders.

There is huge room for improvement and to answer the question if EMPs with their intimate knowledge of the market are focusing on the right things or not, the simple answer would be no. Local companies, who predominantly happen to be manufacturers of generics, with comparatively low entry barriers and intense competition are extremely cost conscious. Hunger for profits and growth graphs towards the sky blurs certain key aspects of quality and safety which otherwise are essential. While on paper, the specifications of these products match the standards set by Pharmacopeia's and other regulators, several processes themselves are in question. Would the advanced markets like the USA or Japan ever accept these goods?

> US FDA in January 2014 imposed a ban on the Toansa facility of Ranbaxy that manufactures APIs and prohibited all products manufactured using API from this facility to be sold to US consumers for failing to comply with cGMP. This is in addition to the ban on three other facilities of Ranbaxy located in Paonta Sahib, Dewas and Mohali. The direct results of these are many like loss of revenue and in the immediate aftermath severe erosion of shareholder value, the growing and unmanageable set of problems eventually led to Daiichi Sankyo deciding to sell its share to India-based Sun Pharma. Now that Ranbaxy is operating as part of Sun, in a culture that Ranbaxy calls home, we have to wait and watch how this evolves for stakeholders as well as shareholders.

Fuelled by patent expiries of blockbusters and key products of innovators, importance is being given to developing copies of these products, entering new markets that are primarily advanced markets, expanding reach and optimizing product costing across emerging markets, however, what is being missed is safety in operations, investments in enhancing quality, strict compliance to regulations, keeping pollution (releasing effluents into atmosphere and water bodies) in check, all of which form the heart of sustainability. When these things are given equal or more importance than growing top and bottom lines, supplying products that provide value beyond price and cost would be a possibility.

It would not be bold to say most companies in emerging markets are time bombs. These companies with thorough knowledge of local economics and industry can do so much more in building sustainable value instead of working towards short-term gains. Regulatory aspects have to be more thoroughly incorporated from development to commercializing and thus play a pivotal role in the companies' performance. Another important point to consider is that harmonisation of regulations across different countries has made progress over the years, however, it is still a long way before universal acceptance of a single set of regulations and their interpretations in varying contexts can be achieved. Though a herculean task, efforts in this direction when pushed forward by stakeholders from industry, government and patient groups would pave the way for sustainability and be a leap in the right direction. Especially with the long product development cycles in the pharmaceuticals and devices industry, there is a greater possibility for regulations to change. In light of this, being aware of cross-market regulations for companies operating or doing business across several countries is the need of the hour.

In June of 2014, EU gave a green light to the Toansa facility of Ranbaxy that the US FDA had earlier in the year prohibited from selling to the US market. Though deficiencies were observed, European regulators expressed satisfaction with the corrective measures put in place after US FDA ban, while the US FDA commented post EU approval that their ban still holds. The European regulatory team opined there is no risk to public health from the deficiencies observed (Clarke 2014) and gave the Toansa facility a pass. In retrospect, two agencies from advanced countries are issuing differing assessments on products arising out of the same manufacturing facility. Could this be due to lack of cohesion in understanding of regulations? This calls for even more need to harmonise regulations globally.

The above-discussed imperatives collectively will drive the industry forward and help commercialize healthcare-related products sustainably.

References

Abott T, John AV. The cost of US pharmaceutical price reductions: a financial simulation model of R&D decisions. Working paper. Cambridge: National Bureau of Economic Research. 2005. http://www.nber.org/papers/w11114.pdf. Accessed 8 June 2014.

Anello E. A framework for good governance in the public pharmaceutical sector. 2008.

Bond E, Saggi K. Compulsory licensing, price controls, and access to patented foreign products. 2012. http://www.wipo.int/edocs/mdocs/mdocs/en/wipo_ip_econ_ge_4_12/wipo_ip_econ_ge_4_12_ref_saggi.pdf. Accessed 10 May 2014.

Burgess CL, Kand E. "The Next Hot Thing"? DOJ makes cGMP enforcement a top priority. 2013. http://www.pharmacompliancemonitor.com/the-next-hot-thing-doj-makes-cgmp-enforcement-a-top-priority/5625/. Accessed 18 June 2014.

Clarke T. Europeans, U.S. differ over safety of Ranbaxy facility. 2014. http://in.reuters.com/article/2014/06/05/ranbaxy-india-idINKBN0EG1PH20140605. Accessed 5 June 2014.

Competition Policy and Public Procurement: Round table on competition policy and public procurement. Geneva: UNCTAD; 2012. pp. 2–3.

Farrell D et al. Accounting for the cost of US health care: a new look at why Americans spend more. 2008. http://www.mckinsey.com/insights/health_systems_and_services/accounting_for_the_cost_of_us_health_care. Accessed 10 June 2014.

Honig P. International harmonization: an industry perspective. International regulatory harmonization amid globalization of drug development. Washington, DC: National Academy; 2013. http://www.ncbi.nlm.nih.gov/books/NBK174222/#sec_011. Accessed 18 May 2014.

Hooper CL. Pharmaceuticals: economics and regulation. 2008. Library of Economics and Liberty. http://www.econlib.org/library/Enc/PharmaceuticalsEconomicsandRegulation.html#IfHendersonCEE2-125_footnote_nt356. Accessed 10 June 2014.

ICH: 2014. http://www.ich.org/about/faqs.html. Accessed 7 April 2014.

Pharmaceutical Price Controls in OECD Countries Implications for U.S. Consumers, Pricing, Research and Development, and Innovation: Washington, DC: U.S. Department of Commerce—International Trade Administration. 2004. http://www.trade.gov/td/health/DrugPricingStudy.pdf. Accessed 28 June 2014.

Reichman JH, Hasenzahl C. Non-voluntary licensing of patented inventions—historical perspective, legal framework under TRIPS, and an overview of the practice in Canada and the USA. 2003. http://www.ictsd.org/downloads/2008/06/cs_reichman_hasenzahl.pdf. Accessed 5 July 2014.

Reichman JH. Compulsory licensing of patented pharmaceutcal Inventions: evaluating the options. J Law Med Ethics. 2009: 258. http://scholarship.law.duke.edu/cgi/viewcontent.cgi?article=2747&context=faculty_scholarship. Accessed 5 July 2014.

Rosenberg T. The fight against fake drugs. 2014. The New York Times. http://opinionator.blogs.nytimes.com/2014/06/04/the-fight-against-fake-drugs/?_php=true&_type=blogs&hp&rref=opinion&_r=0. Accessed 16 June 2014.

Sridhar D, Gostin L. Caring about health. 2010.

Taiwan FDA:. Related announcements. 2013. http://www.fda.gov.tw/tc/siteList.aspx?pn=1&sid=3014. Accessed 30 June 2014.

Top 10 Drugmakers in Emerging Markets: October 2013. http://www.fiercepharma.com. Accessed 15 Feb 2014.

WHO: Medicines: spurious/falsely-labelled/falsified/counterfeit (SFFC) medicines. 2012. http://www.who.int/mediacentre/factsheets/fs275/en/. Accessed 12 June 2014.

Wileman H, Mishra A. Drug lag and key regulatory barriers in the emerging markets. Perspect Clin Res. 2010:51–56.

Chapter 7
Disease Management in the Perspective of Sustainable Growth in Health-Care System: Is Disease Management a Good Business Model for the Sustainability of Health-Care System?

Fereshteh Barei

Introduction

It is noted that innovation is the key to prevention and disease management; this can lead the sustainability of any health-care system. Over the years, innovation has been a driving force in improving treatment results, patient outcomes, and health system efficiency.

Regarding this matter, new technologies have been developed and contributed to treatments. New medicines are providing novel ways to fight disease and maintain patient's health, while new diagnostic tools are helping in earlier detection and more effective management. New information and communication technologies are opening up whole new ways for health-care providers to work together for both better diagnosis and better treatment application.

The "cost" of innovation is the major consideration of policy makers and health system administrators. It is certainly true that an important investment of time, energy, and money is needed to pursue real innovation.

One way to ensure the sustainability of health-care systems is applying an appropriate "business model" to these systems—this will put innovation to work and will also optimize the disease management and patient adherence to the treatment.

Research Methodology

A literature search was performed using MEDLINE, EMBASE, EBSCO, and the Wiley Library, to identify potential studies. The publication search dates varied by review, but typically ranged over 5–10 years of literature (specific details are

---

F. Barei (✉)
LEGOS, Paris Dauphine University, Laboratory of Health Economics and Management of Health Related Organisations, Place du Maréchal de Lattre de Tassigny,
75775 Paris Cedex 16, France
e-mail: fereshteh.barei@gmail.com

available in the individual reports). Abstracts were reviewed by a single reviewer and, for those studies meeting the eligibility criteria, full-text articles were obtained. Reference lists were also examined for any additional relevant studies not identified through the search.

A qualitative method is used to analyze the five interviews that were conducted during this research using Nvivo10 software to identify common ideas.

The Evolution of Disease Management Concept: Better Outcomes at Lower Costs?

The term "disease management" typically refers to multidisciplinary efforts to improve the quality and cost-effectiveness of care for select patients with chronic illness, this trend emphasizes on the importance of assessing the clinical and public policy implications from the perspective of patients' best interests and the quality of care. (Improving quality of care through disease management, principles, and recommendations from the American Heart Association's expert panel on disease management (Faxon 2004). Disease management (DM) is usually applied to diabetes, asthma, heart failure, or depression)

Besides reducing costs, DM aims to improve patient compliance to pharmaceutical drugs. It ranges from educating patients about appropriate self-care to developing customized plans in coordinating care for patients with multiple chronic conditions—some DM programs also try to improve providers' adherence to evidence-based care guidelines.

The Value Proposition of DM Program: Implementing Educational Change

The concept of DM is mainly associated with "knowledge sharing." It is a public health strategy as well as an approach to personal health. The value proposition of DM can be observed in reducing health-care costs and/or improving the quality of life for individuals by preventing or minimizing the effects of disease, usually a chronic condition, through knowledge, skills, enabling a sense of control over life (despite symptoms of disease), and integrative care.

The underlying premise of DM is that *when the right tools, experts, and equipment are applied to a patient population, labor costs (specifically: absenteeism and direct insurance expenses) can be minimized in the near term, or resources can be provided more efficiently*. The general idea is not only facilitating the disease path but also if possible preventing the disease itself. Improving quality and activities for daily living are first and foremost. Improving cost, in some programs, is a necessary component, as well.

DM as a system or tool has undergone evolution. The role of innovative evolution of DM begins when the positive impact of new ways of delivering care are recognized and need to be shared publicly within health-care organizations and through patient populations.

If the adoption of innovation into practice is proven successful, the evaluation is shared with other practitioners to expand the use of the new innovation.

The innovative evolution of DM can help to overcome the increasing pressure of health-care expenditure, using the information emerging from proper cost-effectiveness evaluations can help to develop clear guidelines for improved DM programs (Schwermann and Greiner 2003).

Eventually, the innovative practice can be broadly accepted if it is cost-effective. In its simplest form, dissemination is about sharing what works through interpersonal communication. If those who innovate have limited capacity to share their findings with relevant communities of interest, the innovation will not spread. It is vital, therefore, that those working on introducing innovation on the frontline are given the opportunity to talk about their experience.

Enabling Innovation Adoption

The adoption of innovation happens only when the benefits of implementing change clearly outweigh the costs of change in the previous status. For the diffusion of innovation, clear adoption strategies are needed to make innovation acceptable; this is despite their relative advantages, and, therefore, comes the need for a business model.

If people working in the system clearly understand how particular changes can improve patient care, then the implementation in their own practice is far more likely. Innovations spread faster when frontline providers have a high degree of trust in those communicating the impact of change. (Improving chronic disease management and health system sustainability in Ontario the better care faster coalition, 2013)

Managing Innovation Process for Sustainability of Health-Care Systems

The focus on the way how we identify, support, and disseminate high-value innovation can be considered as a starting point. The benefits of "new innovation" in DM are not always proved or approved across the "health-care consumers" network. Those who study how innovation in health care is propagated, argue that the problem of low adoption cannot be solved by simply creating new government policies.

The organization of innovation and understanding the innovation process can help to accelerate the results or to influence the whole process to get better outcomes.

This is pursued by empirical studies on successful cases and thereby describing how they organize innovation (e.g., Van de Ven and Poole (1990); Rothwell et al. (1974); Andrew et al. (2007).

In DM, identifying sustainable paths to growth is perhaps the best practice model that can help policy makers, insurance providers, and patients/consumers.

The search for sustainable growth includes the search for new health-care technologies. This requires "business models" that make these technologies acceptable to the patient/customer.

In this chapter, we have provided a brief outline of a new framework that highlights how "optimizing" DM *can improve sustainability in health-care systems.*

The sustainability of health-care system is now a common objective for both policy makers and pharmaceutical companies that seek economic interests. The pharmaceutical industry searches for new ways to market prescription drugs. Developing chronic DM programs is indeed one of their strategies (Cave 1995). In doing so, pharmaceutical companies work with clinical leaders or physicians on DM guidelines to reduce practice pattern variations and improve the quality of patient care (Buchanan 2007).

Innovative Business Models for Sustainable Health-Care Systems

Business models are still somehow unknown to health-care providers and users, the definition is rather new. One of the most widely cited definitions, by Amit and Zott (2001), frames business model as "the design of transaction content, structure and governance so as to create value through the exploitation of business opportunities" (Tersgo and Visnjic 2011).

The business model can also be seen to define the structure of the value chain, i.e., the set of activities from raw materials through to the final consumer with value being added throughout the various activities (Amit and Zott 2010).

The set up and organization of business activity of a firm and the way they compete in their market is the base of a *modern business model, but* traditionally this term was used to describe the activity only at *firm level.*

The health-care system is essential and complex to manage within the boundaries of a given country or region. Over a relatively short period of time, health-care sectors have become one of the fastest-growing sectors. Health-care management is no longer limited to the focus on therapies, treatments, and its application, but is also redefining a role to focus on the *prevention measures* and the long-term well-being of the population (Tersago and Visnjic 2011).

Health care as an economic and business consideration is requiring effective policy. It seems necessary to use specific strategies to encourage implementation of research-based recommendations and ensure changes in practice (Bero et al. 1998).

In practice, for presenting a simple understandable business model, Alex Osterwalder's (www.businessmodelgeneration.com/canvas) original nine building blocks, is still an innovative method that can also be used in health care (Osterwalder and Pigneur 2010). This model integrates aspects of Michael Porter's definition of shared value and Clayton Christianson's concept of "jobs to be done" (Clayton et al. 2008).It measures the value of a business model in its feasibility to deliver value, as well as its ability to deliver on the patient health outcome achieved per health-care investment. There is also a community of health care and business model experts that are working together to change the system (Kevin Riley 2013, http://healthmodelinnovation.com/overview-modelh-business-model-canvas-healthcare/).

In their model called model H, a visual language is used for health-care system thinking, problem solving, and solution design. This can enable the managers and decision makers to generate the business models and communicate them across audiences. The model H tries to deliver innovative solutions to the patient by presenting a cocreation business model considering patients' needs and their evolving expectations.

In the new business model, besides the direct role of patients, the adaptation of the existing technology to new needs can provide better outcomes. New technologies can be developed to meet new and higher demands of an evolving and aging population. Understanding the importance of what new technologies offer to advance goals of improved quality and efficiency in health care can help create new business models and may support their deployment (Coye et al. 2009)

The Contribution of Galenic Alternatives to DM: The Case of Polypill[1] and Drug Repositioning

According to a recent research (Bryant et al. 2013), the majority of patients participating in a survey about reducing the number of pills in cardiovascular treatments found the concept of the polypill very attractive (Yusuf et al. 2012).

The benefits of convenience, reduced pill burden, improved safety associated with reduced confusion about dosing, and reduced cost were all key factors that made the polypill favorable. Conversely, participants had concerns with the inflexibility and efficacy of the polypill. Many enquired about dose changes and various formulations of the polypill that would be required for those who needed dose titration. Other concerns were the manufacturer reliability, subsidy issues, and tablet size. Willingness of participants to switch to a combination therapy may be hampered if a polypill formulation that mirrored their current regimen was unavailable.

---

[1] What is a polypill? A polypill is a medicine that is still in the research phase. It is being developed to potentially prevent and treat cardiovascular disease (CVD) and contains several different medicines in one tablet or pill. In 2003, an article about polypills was published in the British Medical Journal. The article attracted major media and public interest. It looked at combining six different medicines in a single tablet (a polypill) and suggested that a polypill could significantly reduce the risk of two forms of CVD: coronary heart disease (CHD) and stroke. The six medicines that a polypill could include are aspirin, a cholesterol-lowering drug called a statin, folic acid, and a low dose of three blood pressure-lowering medicines, a diuretic, a beta-blocker, and an angiotensin converting enzyme inhibitor. (http://www.heartfoundation.org.au/SiteCollectionDocuments/A-PolyPill-QA.pdf)

Saving Time and Money

Drug repositioning or repurposing is another strategy that can save time and money for health-care consumers and health-care systems. It is a strategy by which new or additional value is generated from a drug by targeting diseases other than those for which it was originally intended (Thomson Reuter report 2012, http://thomsonreuters.com/business-unit/science/subsector/pdf/knowledge-based-drug-repositioning-to-drive-rd-productivity.pdf).

Pharmaceutical products that have been developed and approved for one disease may be the object of additional clinical development in other disease areas or of additional pharmaceutical development for new and different formulations. The newly developed products can be named as repositioned or reformulated products (Murteira 2014).

Saving time and delivering fast access to the high-quality treatment is an important investment in health-care systems. Drug repositioning is generally a faster process than new development because it can rely on existing data. "For a company that takes drugs from target discovery to the market, developing an NCE (new chemical entity) can take 10 to 17 years, depending on indication. For a drug repositioning company, the development process from compound identification to launch can be around 3 to 12 years" (David Cavalla, Ph.D., founder of Numedicus; Eldvige 2010).

The amount of time saved will depend on the amount of data availability, the stage of its development, and the length of trials required. The time taken to develop a repositioned drug also will depend on the indication, as an example a new chronic indication will take longer than simply improving a drug's efficacy, safety, ease of dosing, or dosing frequency in the original indication (Eldvige 2010).

Enhancing Patient Convenience

"Patient-centered care" as care that is "respectful of and responsive to *individual* patient preferences, needs, and values and ensur[es] that patient values guide all clinical decisions (Supporting Patients' *Decision-Making* Abilities and Preferences, www.ncbi.nlm.nih.gov › NCBI › Institute of Medicine (USA) Committee on Crossing the Quality Chasm: Adaptation to Mental Health and Addictive Disorders. Washington (DC): National Academies Press (USA); 2006).

The patient's convenience can be enhanced by many of the approaches. For instance, less frequent dosage or noninvasive delivery can be expected to improve compliance.

The importance of the oral route of administration (Pareek 2010) from both a clinician and patient acceptance point of view means there has been a vast amount of development and research in drug delivery via this route. Noncompliance can be attributed to poor taste, difficulty in administration or swallowing, and the inconvenience of multiple doses per day.

Controlled delivery products are an example of how an innovative drug delivery technology has enabled the development of more convenient dosing regimens that improve patient compliance. This type of product is now routinely used in life-cycle management (Ross 2010).

The time gain, plus the reduction in attrition because of fewer safety concerns compared to a new drug, makes drug repositioning a sensible approach for patient advocacy groups, patients, and their caregivers. An example would be the Michael J. Fox Foundation, whose recent request for funding proposals includes repositioning as a source of novel Parkinson's disease drugs. [2]

Beside the advantages of drug repositioning for patients' convenience, it can open up new markets[3].

Patient's Adherence

Patient's "adherence" to using their treatments is becoming a real economic and business issue. This is also a challenge for DM optimization.

The question is how to encourage and educate the patients to continue their treatments? Nowadays, patients/consumers have access to the information on Internet, and they can choose their *preferred treatment*. Other health-care system stakeholders, such as payers, institute managers, and nurse practitioners, are also taking an increasing role in treatment choices. Within the new marketing methods, we can name digital channels and user online communications as some of the popular and often used by patients—here, they think that *the voice of normal customers* is probably heard (http://blog.gfk.com/2013/11/healthcare-which-marketing-channels-have-most-impact/). Feedback and recommendation from colleagues are also shown to have a strong influence on physicians' prescription.

The *two most commonly identified drug therapy problems in patients receiving comprehensive medication management services are: (1) the patient requires additional drug therapy for prevention, synergistic, or palliative care; and (2) the drug dosages need to be titrated to achieve therapeutic levels that reach the intended therapy goals.*[4]

Many people find it difficult to take their medications and the number of medicines may seem overwhelming. Multiple medications and complexity of treatment regimens are major determinants of poor medication adherence (Brycent et al. 2013).

Providing a framework for integrating comprehensive medication management is also required in any health-care system.

---

[2] From: Integrating comprehensive medicine management to optimize patient outcome, http://c.ymcdn.com/sites/www.chronicdisease.org/resource/resmgr/cvh/integrating_comprehensive_me.pdf. By ArisPersidis, PhD, President; Biovista, Inc., Charlottesville, Va., "Drug Repositioning: A Union of Patient Interests, Pipeline Development and Innovation", http://biovista.com/drug-repositioning-a-union-of-patient-interests-pipeline-development-and-innovation/, 2011.

[3] Teaching Old pills new tricks, by A.R/Oxford, www.economist.com, Feb 2013.

[4] Integrating comprehensive medication management to optimize patient outcomes, June 2012. http://c.ymcdn.com/sites/www.chronicdisease.org/resource/resmgr/cvh/integrating_comprehensive_me.pdf

The processes of writing and filling a prescription are important elements of using pharmaceutical treatments, the educational systems do not present the technical aspects of these activities precisely everywhere. It should be clear that both activities need to occur with accuracy for patients to be well defined and served.

The assessment begins by studying the patient's medication experience—that is concerning the identification of patient's beliefs, understanding, and expectations about his or her medications. This helps to understand patient decision making about (a) whether to have a prescription filled, (b) whether to take it, (c) how to take it, and (d) how long to take it. The goal of medication management is to positively impact the health outcomes of the patient, which necessitates actively engaging them in the decision-making process. Therefore, it is essential to understand the patient's medication experience (Griesbach 2012).

Further scientific research is needed to explore the validity of patients' experiences as a progression through stages of treatment. "This is essential for health care practitioners and disease management providers to acknowledge an individual's medication experience in order to positively influence medication taking behaviours. Patients' decisions, which at first may appear irrational, might be seen as intelligent when a practitioner, prescriber understands a patient's unique medication experience" (Shoemaker and Dienane 2008). Without a good understanding of the patient's medication experience, sound clinical decisions cannot be taken (Cipolle et al. 2012).

The role of social media and professional blogs and websites like PatientsLikeMe in the USA, Carenity in France, etc., can help patients to share their experience and ask their questions. They may also contribute to surveys and provide a source of information for decision makers for preventing and resolving drug therapy problems.

Implementation of a "Well-Prepared Medication Delivery" to Patient: The Delivery of Comprehensive Medication Management

Health professionals that possess the knowledge of medication management have an understanding of the comprehensive taxonomy of drug therapy problems, and the ability to apply the rational and systematic decision-making process for drug therapy. (Supporting Patients' Decision-Making Abilities and Preferences, http://www.ncbi.nlm.nih.gov/books/NBK19831/)

The current academic preparation of pharmacists *should* qualify them to deliver medication management services to patients. An additional training may be required to meet the national standards. Many pharmacists can provide this service.

Education and using pharmacists who can manage difficult, complex patients, or elderly patients with medication problems will make the entire patient care team more effective and efficient. Hospitals as well as pharmacies are now meeting the consequences of problems associated with lack of knowledge or misuses of medication and drug-related morbidity and mortality; this can be changed by a better disease management. Medication management optimizes drug therapy not only in ordinary patients but also in patients who need additional time and attention; this should result in better management of health-care costs everywhere (Street et al. 2009).

DM can also produce better clinical results in several cases such as diabetes and hypercholesterolemia.

All therapeutic outcomes can be improved by using medication management. By identifying goals, all *medications can be assessed, drug therapy problems can be identified and solved,* and actual outcomes can be continuously evaluated until appropriate outcomes are achieved (The Patient-Centered Medical Home: Integrating Comprehensive Medication Management to Optimize Patient Outcomes RESOURCE GUIDE second edition June 2012)

Health IT Adoption, Patient Data and Disease Management

An initial question for every health organization in implementing projects for DM is whether to buy a commercially available health IT product or not? Regardless of selected approach, a significant investment of time and resources is required to configure IT systems to perform the functions desired not only by consumers but also by stakeholders (From: *Health IT for improved Chronic disease Management,* http://healthit.ahrq.gov/ahrq-funded-projects/emerging-lessons/health-it-improved-chronic-disease-management)

Securing user buy-in and trust is critical to the success of health IT implementations in DM.

More technical aspects and equipment do not always result in better care. This is particularly true for applications developed for patient use, such as integrated voice response (http://healthit.ahrq.gov/ahrq-funded-projects/emerging-lessons/health-it-improved-chronic-disease-management) and patient portals, which can become so complex to use for the patients specially if they are old and this will discourage user adoption. It is also noticed that it was important to keep the user interfaces and options as simple as possible.

Short-term health IT solutions may be put in place to respond to the immediate needs. These systems can be implemented rapidly and adopted by clinicians. At the same time, a support system is needed to be developed and tested. The project can be a working data exchange that enables providers to easily refer patients and receive feedback on referral encounters. Once fully used, other aspects of clinical data exchange can facilitate sharing of additional forms of clinical data.

Facilitating Collaboration Between Patients and Providers, Increasing Quality of Care

Health IT can enable opportunities for remote patient management, patient education, and provider information sharing for patients with chronic conditions (Shapiro and Barton 2013).

Providing up-to-date access to information for patients about clinical practice, medications, and treatment options is very important. Some examples of the ways that these future methods are used to educate patients and providers are described below:

Telehealth[5] network helps physicians receive up-to-date information about clinical practices for chronic conditions. Physicians also can interact with other primary care physicians and specialists at the closest academic medical center to discuss complex cases. According to Darkins, in his case study (2008), this enables physicians to learn from one another. The telehealth network also educates nurses and office managers about processes for teaching patients about self-management of their chronic illnesses. Telehealth also can be used to provide education *directly to patients*.

Studies have shown that there is no difference in the ability of the provider to obtain clinical information, make an accurate diagnosis, and develop a treatment plan that produces the same desired clinical outcomes as compared to in-person care when used appropriately (http://www.americantelemed.org/docs/default-source/policy/examples-of-research-outcomes-telemedicine%27s-impact-on-healthcare-cost-and-quality.pdf).

Improving care for chronically ill patients provides benefits for patients and the whole society. The objective of several projects is achieving sustainability for *chronic care* initiatives by securing support from governmental organizations. The need for continued support for innovative uses of health IT for disease management is inevitable; and with some hope, these interventions can also target the populations that are the sickest and the neediest.

Patient satisfaction from the quality of care by using Telehealth is analyzed in several studies. The use of telemedicine to access care and the use of telecommunications technologies to connect with specialists and other health-care providers in order to meet unmet medical needs have consistently been very high. Degrees of satisfaction may vary slightly with the specialty accessed through telemedicine, but overall patients have responded well to its use. The source of satisfaction for most patients is the ability to see a specialist trained in the area most closely related to the patient's condition, the feeling of getting personalized care from a provider who has the patient's interest in mind, and the ability to communicate with the provider in a very personal and intimate manner over the telecommunications technologies (Gustke et al. 2000).

As for the cost-effectiveness dimension of telehealth a recent study (Gustam et al. 2014), *showed that:* Data on telehealth investment costs are lacking in many studies, but few studies that assessed costs and consequences comprehensively showed

---

[5] Telehealth is the remote provision of healthcare services and health education, mediated by technology. You may hear other similar terms such as telemedicine, e-health, connected health and health telematics—we consider the differences between these terms to be so minor that they are essentially equivalent. The important common thread between all of these terms is that technology is used to break down barriers of geography and access to health care and education.from: http://telehealth.med.miami.edu/what-is-telehealth

that telehealth interventions are cost saving with slight improvement in effectiveness, or comparably effective with similar cost to usual care.

IT technology used in DM meets several challenges. The cost-effectiveness of this advanced "tool" is the major subject in economic evaluation. The increasing quality of care proposed to the patient, patient satisfaction, and patient adoption of this method are very important topics in DM-related studies. The positive effect of telehealth technology can influence positively the sustainability of health-care system.

Evolving Health Care for Quality Care

Many different groups in health-care systems are concerned about quality care. They have to work towards the goal of improving patient care and ensuring that care system will get the maximum return on its health-care expenditures. (Putting innovation to work 2013, http://www.bettercarefaster.ca/BFCPuttingInnovationto-Work.pdf)

The priority is to address one of the biggest health challenges and largest cost drivers in the health-care system—the growing prevalence of chronic diseases. *Our aging population will undoubtedly add to the medical cost crisis and will need more financial resources.*

The key to tackling this enormous societal problem is to begin taking concerted action to ensure that innovative breakthroughs in chronic disease management are rapidly identified and implemented. The answer to improving the efficiency and quality of health-care system lies in early adoption of innovative initiatives that have been proven to make a significant difference in health outcomes. Adopting novel approaches to DM stands to benefit everyone. Patients will have access to timely, equitable, and high-quality health care, leading to better management of their condition, and, consequently, less reliance on health resources. *Finally, the government will gain by having a more efficient health-care system that will be sustainable over the long term* (Marchildon and DiMatteo 2011).

While several innovations exist to help improve efficiency and health outcomes, the problem is that successful health technologies, service models, and practice guidelines are not rapidly adopted everywhere. This is despite evidence of their proven benefit to improve health outcomes and system efficiency. The objective must be creating an environment and promising new innovations identified and adopted (Buntin et al. 2011).

Another problematic strategy to deal with the current health-care funding challenge would be to try shifting the cost of care from the public system to private individuals.

In many cases, it may make more long-term economic sense, in terms of fewer hospital admissions, and health complications.

Recommendations

All health-care systems are struggling to identify and disseminate innovation in a more systematic way.

Meaningful change in health-care system can happen only if health-care institutions, providers, researchers, industry, as well as patients and their families *work together* to find novel solutions. Through concerted effort, the policies can be reoriented, and most importantly our culture, so that innovation is recognized as being critical to improving

Standardize the Evaluation Framework for Innovative Initiatives

Those willing to invest time, money, and expertise in piloting new approaches to care do so with a reasonable expectation that the programs will be adopted into national health-care system.

It is often difficult to convince the Ministry of Health and other health-care providers of the value of an innovation.

The aim to provide quality care and faster care can invest time and resources to test innovative chronic disease management initiatives designed to improve population health and system performance. This will bring meaningful improvements to patients and the health system overall.

Issues to Monitor in the Future

Weaker prospects for economic growth combined with fiscal deficits and fewer savings from debt service charges could have a dampening effect on the future growth of health spending.

As the percentage of the population aged 80 and older increases, decision makers will be faced with the challenge of determining the best ways to provide care for older adults. The challenge will be to find the appropriate use of hospital care, long-term institutional care, and community care for patients that balance access, quality, and appropriateness of care on the one hand and cost on the other.

Price inflation is also a significant cost driver. Managing health-specific price inflation for core health goods and services, including doctors, nurses, other health-care professionals and advanced diagnostics, will be a challenge.

A rapid increase in physician remuneration places considerable cost pressures on all governments.

High-value innovations, when adopted early and consistently across the system, can have a tremendous positive impact on quality of care, patient well-being, and health system economics. It is clear that the impediment is not science but rather lack of opportunity or collective will to work together in concert to take more deliberate action to effect change.

Better and faster care may focus on:

- Engaging scientists, opinion leaders, and the patients and families that shape our health system.
- Communicating with government and government agencies who play a critical role in health policy development, regulation, and oversight.
- Mobilizing and growing our broad coalition of members to ensure progress in this movement to advance change.
- Utilize tools and resources to get their products and solutions to market faster, decrease the sales cycle, and win more profitable business.
- Work with industry experts to build customer-driven technologies, while receiving assistance in expanding into new markets.
- Access leading edge support from the first encounter with the customer to post-sales support.

Concluding Remarks

In this chapter, we mainly focused on the role of innovation-driven methods in DM, the knowledge circulation, the adaptation of innovation in practice, the role of IT, and the galenic contribution to DM and sustainability of health-care systems. According to Drummond, "The ideal health system would put more emphasis on preventing poor health. It would be patient-centric and would feature coordination along the complete continuum, of care the patient may require. Primary care would be the main point of patient contact, with a good part of the coordination across care taking place through the administration of hospitals or regional health authorities" (Drummond 2011).

The aim of the innovation-driven DM programs is providing the optimized care to patients and preserving the sustainability of health-care systems. This may consider many determinants, such as long-term in-home care for their cost-effectiveness, continuity of care, use of new technologies, and the use of improved pharmaceutical treatments like polypills and repositioned drugs to decrease side effects and increase patient convenience. If in this approach patients' unmet needs are satisfied, the new methods will be adopted and patients' confidence will facilitate the whole evolution. Detailed studies of patient perception of treatment, and DM in the case of chronic diseases, is necessary in constructing the excellence in health-care programs.

References

Amit R, Zott C. Value creation in e-business. Strategic Management J. 2001;22(6/7):493–520.
Amit R, Zott C. Business model innovation: Creating value in times of change, Navarra, Spain: IESE Business School Working Paper No. 870. 2010.
Andrew JP, Sirkin HL, Haanaes K, Michael DC. "Measuring Innovation, 2007 – A BCG Senior Managemnt Survey". 2007. Accessed Sept 2013.

Bero LA, Grill R, Grimshaw JM et al. Closing the gap between research and practice: an overview of systematic reviews of interventions to promote the implementation of research findings. BMJ. 1998;317:465–8. Accessed Sept 2013.

Bernstein J, Firlik KS, Chollet D, Peterson G. Disease management: does it work?. N°4, Mathematica Policy Research Inc. Issue Brief. 2010

Bryant L, Martini N, Chan J, Chang L, Marmoush A, Robinson B, et al. Could the polypill improve adherence? The patient perspective. J Prim Health Care. 2013;5(1):28–35.

Buchanan D, Fitzgerald L, Ketly D. The sustainability and spread of organizational change: modernizing healthcare. New York: Routledge, 2007.

Buntin M, Burke M, Hoaglin M, Blumenthal D. The benefits of health information technology: a review of the recent literature shows predominantly positive results. Health Aff. 2011;30(3):464–71.

Clayton CM, Michael BH, Curtis WJ. *"Disrupting Class: How Disruptive Innovation Will Change the Way the World Learns"*. 2008. McGraw-Hill.

Cave DG. Capitated chronic disease management programs: a new market for pharmaceutical companies. Benefits Q. 1995;11(3):6–23.

Chew DP, Carter R, Rankin B, Boyden A, Egan H. Cost effectiveness of a general practice chronic disease management plan for coronary disease in Australia. Australian Health Rev. 2010;34(2):162–9.

Cipolle R, Strand L, Morley P. Pharmaceutical care practice: the patient-centered approach to medication management. New York: McGraw-Hill, 2012.

Coye MJ, Haselkorn A, DeMello S. Remote patient management: technology-enabled innovation and evolving business models for chronic disease care. Health Aff. 2009;28(1):126

Darkins A, Ryan P, Kobb R, et al. Care coordination/home telehealth: the systematic implementation of health informatics, home telehealth, and disease management to support the care of veteran patients with chronic conditions. Telemed J E Health 2008;14:1118–26.

Drummond D. Commission on the reform of Ontario's public health service. 2011. Chapter 5. www.fin.gov.on.ca/en/reformcommission/chapters/ch5.html.

Elliot R, Ross-Degnan D, Adams A, Safran D, Soumerai S. Strategies for coping in a complex world: adherence behaviour among adults with chronic illness. J Gen Intern Med. 2007;22(6):805–10.

Eldvige S. Getting the drug repositioning genie out of bottle. 2010. http://www.lifescienceleader.com/doc/getting-the-drug-repositioning-genie-out-of-the-bottle-0001.

Faxon DP, Fuster V, Libby P. Atherosclerotic Vascular Disease Conference: Writing Group III: pathophysiology. Circulation. 2004;109:2617–2625.

Goodyear Smith F. Patient and provider participation in healthcare provision. J Prim Health Care. 2013;1(5)266–7. https://rnzcgp.org.nz/assets/documents/Publications/JPHC/March-2013/JPHCEditorialMarch2013.pdf.

Griesbach S. Panel discussion on patient engagement. Marshfield Clinic, Marshfield. December 4, 2012

Gustam AS, Severens JL, van Nijnatten J, Koymans R, Vrijhoef HJM. Cost effectiveness of telehealth interventions for chronic heart failure: a literature review. International J Technol Assess Health Care. 2014;30(1):59–68.

Gustke SS, Balch DC, West VL, Rogers LO. Patient satisfaction with telemedicine. Patient satisfaction was examined in relation to patient age, gender, race, income, education, and insurance. Overall patient satisfaction was found to be 98.3 %. Telemed J Spring. 2000;6(1):5–13.

"Heath IT for improved Chronic disease Management", healthit.ahrq.gov/ahrq-funded-projects/emerging-lessons/health-it-improved-chronicdisease-management

"Integration comprehensive medication management to optimize patient outcomes", Resource Guide, second edition, June 2012, The patient-centered Medical home publication, https://www.accp.com/docs/positions/misc/CMM%20Resource%20Guide.pdf

Lamel S, Chambers CJ, Ratnarathorn M, Armstrong AW. Impact of live interactive teledermatology on diagnosis, disease management, and clinical outcomes. Arch Dermato. 2012;148(1):61–5.

Marchildon G, DiMatteo L. Health care cost drivers: the facts. 2011. https://secure.cihi.ca/free_products/health_care_cost_drivers_the_facts_en.pdf

Mays GP, Au M, Claxton G. MARKETWATCH: convergence and dissonance: evolution in private-sector. Health Aff. 2007;26(6)1683–91.

Murteira S, Millier A, Ghezaiel Z, Lamure M. Drug reformulations and repositioning in the pharmaceutical industry and their impact on market access: regulatory implications. J Mark Access Health Policy. 2014;2:22813. http://www.jmahp.net/index.php/jmahp/article/view/22813.

Opera TI, Mestres J. Drug repurposing: far beyond new targets for old drugs. AAPS J. 2012,14(4):759–63.

Osterwalder A, Pigneur Y. "Business model generation", Wiley publication. 2010. ISBN: 978-0-470-87641-1.

Pareek A. Leveraging Innovation in Generics, a business imperative. First conference on Super Generic Innovation, Mumbai, India. 2010.

Putting innovation to work, improving chronic disease management and health systems sustainability in Ontario, a publication from: The Better care Faster Coalition. 2013. http://www.bettercarefaster.ca/BFCPuttingInnovationtoWork.pdf. Accessed Sept 2013.

Riley k. "Creating a culture of innovation for health plans", Slide share presentation. 2013. http://www. slideshare.net/kevineriley/2013-03-creating-a-culture-of-innovation-for-health-plans

Ross MS. Innovation strategies for generic drug companies: moving into supergenerics. IDrugs. 2010;13(4):243–7.

Rothwell R, Freeman C, Jervis P, Robertson A, Townsend J. SAPPHO Updated: Project SAPPHO Phase 2. Research Policy. 1974;3(3):258–291. Accessed Sept 2013.

Schwermann T, Greiner W. Value in health, vol 6, supp 1. "Using disease management and market reform, to address the adverse economic effects of drug budgets and price and reimbursement regulations in Germany. 2003.

Shapiro Mathews E, Barton AJ. Using the patient engagement framework to develop an institutional mobile health strategy. Clin Nurse Spec. 2013;27(5):221–3

Shoemaker JS, Dienane R. "Understanding the meaning of medications for patients: The medication experience". Pharm World Sci. 2008;30(1):86–91. PMCID: PMC2082655. Accessed Sept 2013.

Street RL Jr, Makoul G, Arora NK, Epstein RM. How does communication heal? Pathways linking clinician–patient communication to health outcomes, Elsevier. 2009;74(3):295–301.

Supporting Patients Decision-Making Abilities and Preferences. www.ncbi.nlm.nih.gov › NCBI › Institute of Medicine (US) committee on crossing the quality chasm: adaptation to mental health and addictive disorders. Washington (DC): National Academies Press, 2006.

"Teaching old pills new tricks" a Schumpeter Business column , The Economist journal , Published februray 2013, http://thomsonreuters.com/business-unit/science/pdf/ls/drug_repositioning-cwp-en.pdf

"Telemedicine's Impact on health care costand quality, 2013, American Telemedicine Association, http://www.americantelemed.org/docs/defaultsource/ policy/examples-of-research-outcomes-telemedicine%27s-impact-on-healthcare-cost-andquality. pdf

Tersago S, Visnjic I. Business model innovations in health care. Cambridge service alliance.org. 2011.

The patient-centered medical home integrating comprehensive medication management to optimize patient outcomes, 2nd edn. June 2012. http://www.pcpcc.org/sites/default/files/media/medmanagement.pdf.

Thomson Reuters white paper report "Knowledge based drug repositioning to Drive R&D". 2012. http://thomsonreuters.com/businessunit/science/pdf/ls/drug_repositioning-cwp-en.pdf

Van de Ven AH, Poole MS. "Methods for studying innovation. Development in Minnesota Research Program," Organization Science. 1990;1(1)99:313–333. Accessed Sept 2013.

Yusuf S, Pais P, Sigamani A, et al. Comparison of risk factor reduction and tolerability of a full-dose polypill (with potassium) versus low-dose polypill (polycap) in individuals at high risk of cardiovascular diseases: the second Indian Polycap Study (TIPS-2) Investigators. Circ Cardiovasc Qual Outcomes 2012;5:463–71.

Zickmund SL, Hess R, Bryce CL, McTigue K, et al. Interest in the use of computerized patient portals: role of provider-patient relationship. J Gen Intern Med. 2008;23 (1):20–6.

Further Readings

Car J, Black A, Anandan C, Cresswell K, Pagliari C, McKinstry B, Procter R, Majeed A, Sheikh A. The impact of health on the quality and safety of healthcare, report for NHS connecting for health evaluation programme. London: Imperial College, 2008.

Edin M. Traditional tools can be applied to specialty pharmacy management. 2011, pp 34–7.

Elley CA. Polypill is the solution to the pharmacological management of cardiovascular risk: yes. J Prim Health Care. 2009;1(3):232–4.

Firlik KS. Motivating successful disease self-management. Health Manag Technol. 2011;32(12):32.

Sanz G, Fuster V. Fixed-dose combination therapy and secondary cardiovascular prevention: rationale, selection of drugs and target population. Nature Clin Pract Cardiovasc Med. 2008;6(2):101–10.

Yusuf S. Combination pharmacotherapy to prevent cardiovascular disease: present status and challenges. Eur Heart J. 2013;34(13):945–6 (Population Health Research Institute, McMaster University and Hamilton Health Sciences, Ontario).

Chapter 8
Thoughts on Sustainable Health Care…in a Patient-Centric Society

Virgil Simons

The concept of "sustainability" immediately begs the question of "What are we sustaining?" As patients, the government defines us as "consumers," of what always made me ask, "When did I ask for a pound of prostate cancer and go light on the impotence." However, I suppose the argument can be made that we consume health entitlements through Medicaid, Medicare, Social Security, and programs through federally funded education, research, and treatment initiatives. But do these activities devolve from the patient needs analysis or from the establishment's interpretation of, and provisions for, it? And, at what level should they be sustained?

The Institute of Medicine (IOM) defines patient-centered care as: "Providing care that is respectful of and responsive to individual patient preferences, needs, and values, and ensuring that patient values guide all clinical decisions (Institute on Medicine)."

Given that nonconsumer stakeholders (medical centers, pharmaceutical manufacturers, public health agencies, and health-care professionals) often do not know what matters most to patients regarding their ability to get and stay well (Sepucha et al. 2008), care that is truly patient-centered cannot be achieved without active, and ongoing, patient engagement at every level of care design and implementation.

The concept of greater patient involvement in health-care delivery and design (Jo Anne et al.) is driving much of the conversation relative to how we understand and access the care that we need, from pharma company collaborative awareness campaigns, to Accountable Care medical practices, to multi-disease site public health initiatives. Many of the barriers to this objective stem from insufficiency in patient risk awareness and/or disease education coupled with systemic problems in delivering appropriate access to health care. A classic example can be seen in the disproportionate incidence and mortality of prostate cancer within African-American men in the USA.

V. Simons (✉)
The Prostate Net, Passeig de Gracia, 115 Piso 4, PTA 3, 08008 Barcelona, Spain
e-mail: virgil@prostatenet.org

© Springer International Publishing Switzerland 2015
P. A. Morgon (ed.), *Sustainable Development for the Healthcare Industry*, Perspectives on Sustainable Growth, DOI 10.1007/978-3-319-12526-8_8

Many hypotheses have been put forward as to the causality of this situation; however, while scientific investigation has shown some factors of genesis, important progress has been made through increased access to information resulting in early detection and treatment. But this has not been an easily achieved objective due to socioeconomic factors relating to access to, and cost of, care, complicated by historic mistrust of the medical establishment tracing back to the Tuskegee experiments. Clearly, a nontraditional intervention was necessary.

In 2005, The Prostate Net launched its Barbershop Initiative, in partnership with MGM Studios and several regional medical centers and public health agencies, to utilize the release of the movie *Barbershop 2* to promote disease risk awareness and early access to care among the target populations. As a result, more than 30,000 men were screened for prostate cancer that had not been engaged with the medical system prior. The program continues today as an extension of various state comprehensive cancer control programs to drive ongoing education and detection services not only for prostate cancer but also for colorectal cancer and correlative men's health issues.

The core element of sustainability for this initiative has been the active engagement by community barbers, working with local medical centers, to provide critical information, access to centers of care, and motivation to participate in their clients' personal health responsibilities. Pharma partnerships were the key in the launch and advancement of this initiative in 2005 in part because of the uniqueness of the program and the potential media exposure to be generated. However, as the program matured, and negative perspectives arose as to the viability of prostate-specific antigen (PSA) screening, coupled with pharma prioritization of advanced-stage disease therapeutics, we have seen that the involvement goes to zero.

We see much of this new mandate for engagement in the inclusion of patient advocates as part of research advisory panels, institutional review boards (IRBs), industry–advocate collaborative groups, community oversight councils, and the like. It engenders much promise that the bad old days of patients without power are gone. But, are they? Have we really arrived to a point where patient centricity is a fact or is it still a wished-for illusion?

In the ideal world of patient-centered care and multidisciplinary engagement, the patient would have his/her clinical status reviewed collectively by physicians from each of the potential disciplines of therapeutic care and a collective decision taken, with the patient's input, as to the one with best outcomes therapeutically and quality of life. True centricity, but far from the reality of most patients when physicians' performance is evaluated by RBU production and revenue-oriented decision making. Is the issue of sustainability related to patient care and the quality of life or to that of the fiscal health of the institution?

While there have not been any RCT studies to compare systems and protocols, inferences can be drawn between the California University medical system wherein doctors are on fixed salaries versus those where compensation is determined by number of patients seen, number of diagnostics ordered up, and number of procedures performed. We can look at the National Health Service of the UK where a more relevant incentive system is in place based on quantitative improvement in the patient's health.

The Accountable Care Act in the USA seeks to address some of these issues by encouraging amalgamation of primary and some specialty care services into groups that should be able to deliver higher standards of care at lower costs. This hypothesis still needs to be proved because effectiveness and cost reductions are components of medical reimbursement either through private insurers or government agencies.

As advocates we sit in these sessions, advise on these panels, suggest new protocols, recommend new therapies; but are we really changing anything. We revel in our accomplishments as "partners" with pharma in meeting the needs of our patient constituencies. Yet, despite their outward manifestations of support and collaborative spirit, there continues to exist, to one degree or another, depending on the global space, that patients are something to be dealt with as revenue sources, markets to be exploited, and the like, hardly to be embraced as an equal partner in meeting corporate, community, and patient-centric needs.

True sustainable patient-centered care must begin before care is necessary. Consumers must engage with the legislative and health establishment to set priorities for government expenditures on preventive education and intervention, research funding priorities, restructuring of government payments for therapeutic care, prioritizing pharmaceutical drug development, and creating healthier environments.

We see today the approval and utilization of many new therapies for advanced-stage cancers, but with extremely high prices and limited patient effectiveness. As a nation, are we better serving our citizens by paying with our tax dollars for an agent that costs almost US$100,000, is only effective in approximately 30% of the affected patient populations, and offers only a few months of survival. Would not our return-on-health investment be more effective in using those dollars to promote better health behaviors and early detection for a healthier society?

We need to analyze the cost of advanced-stage disease care versus that of proactive prevention and invest in initiatives that will stem the increase in chronic diseases of lifestyle, e.g., diabetes, obesity, etc., while concurrently increasing our support for research and information sharing towards heightened opportunities for cure and chronic care management.

And we need to break down the barriers to true universal access to care to insure that there will be true equitable sharing in the basic human right to good health and quality of life.

If we cannot become more educated and proactive consumers, if we do not embrace our responsibilities to ourselves and our communities relative to informed choice, and if we allow our "centricity" to be determined by special interests other than ours, if we do not define the real parameters of sustainable health care, we will see a continued erosion of those resources we value, and need, in profligate and unfocused consumption.

References

Institute on Medicine. Crossing the quality chasm: a new health system for the 21st century. Accessed: 26 Nov 2012.

Jo Anne LE, Elizabeth AF, Melissa BG. Patient advocacy for health care quality.

Sepucha KR, Levin CA, Uzogara EE, Barry MJ, O'Connor AM, Mulley AG. Developing instruments to measure the quality of decisions: early results for a set of symptom-driven decisions. Patient Educ Couns. 2008;73(3):504–10.

Chapter 9
The Biopharmaceutical Industry is Part of the Solution for Healthier, Wealthier Societies: Interview with Dr. Eduardo Pisani

Pierre A. Morgon

The insights stemming from the discussion with Eduardo Pisani, Director General of IFPMA[1], help put the sustainable development questions in the global perspective, especially by underscoring the pivotal value of global partnerships in the virtuous circle of reinforcing both health and wealth, and the need for open collaboration and dialogue between all stakeholders.

Pierre Morgon: The context is the preparation of a book focusing on sustainable development in the life-science industry, assembling inputs from the vantage points of the various stakeholders. The core topic is to address what the industry should do differently so as to retain its role as a pivotal contributor to society and global health and wealth. From the IFPMA perspective, could you comment on what has been done and what should be done in this respect?

Dr. Eduardo Pisani: Sustainable development is indeed a crucial concept and a key topic at present as it relates notably to the post-2015 sustainable development goals of the United Nations. The

---

[1] IFPMA represents the research-based pharmaceutical companies and associations across the globe. The research-based pharmaceutical industry's 1.3 million employees research, develop, and provide medicines and vaccines that improve the life of patients worldwide. Based in Geneva, IFPMA has official relations with the United Nations and contributes industry expertise to help the global health community find solutions that improve global health.

P. A. Morgon (✉)
Theradiag, Croissy-Beaubourg, France
e-mail: pierre.morgon@wanadoo.fr; mrgn@bluewin.ch

Eurocine Vaccines, Solna, Sweden

AJ Biologics, Kuala Lumpur, Malaysia

© Springer International Publishing Switzerland 2015
P. A. Morgon (ed.), *Sustainable Development for the Healthcare Industry*,
Perspectives on Sustainable Growth, DOI 10.1007/978-3-319-12526-8_9

formulation of such goals offers the entire health-care industry an unprecedented opportunity to contribute to their achievement. Why so? Because each stakeholder, including us, will have a chance to help making the world more prosperous, fair, and sustainable. All this must ultimately fall within a holistic vision on what a society should be. We are talking about social inclusiveness, economic yields that are widely shared, equality, and shared prosperity. In essence, these goals for sustainable development are not new, as they stem from a continuum based on the Millennium Development Goals that have been set some 15 years ago. *From a global health policy perspective, there is time for an even greater ambition*, the more so as the world population is expected to grow to 8 billion in a decade. So you can imagine that also the industry will have complex targets for itself in order to help tackle challenges that will affect us all: ageing population, increase of chronic and non-communicable conditions like cancers, diabetes, lung and heart diseases, as well as the spread of infectious diseases in expanding urban areas, to name just a few of the irrefutable trends[2]. The health-care industry will of course have to play an important role to contribute to global sustainable development goals. If we look at the three dimensions of the matter: how the social, economic, environmental dimensions interact with each other to generate sustainable development, we find that they all aim at one common goal—human development. If I break down these three dimensions, one can immediately see the key role of the biopharmaceutical industry. Let me take the social dimension, for instance. First, there is a systemic component. This means that contributing to the establishment of robust health-care systems provides the appropriate level of care to all people around the world. We, as an industry, contribute to strengthening health-care systems for instance through capacity building efforts, educational activities, public awareness campaigns, and of course a number of other investments, which not only are related to economic factors but also are truly focused on societal benefits. Innovative vaccines for instance have generated tremendous value by preventing disease and sustaining healthy communities.[3]

---

[2] http://www.thelancet.com/commissions/global-health-2035.

[3] http://www.ifpma.org/fileadmin/content/Publication/2014/value_of_innovation.pdf.

PM: **Based on previous experience in vaccines, I have been involved in a number of educational activities and infrastructure strengthening, duly recognized as contributing to human development.**

Dr. Pisani: You are right. Prevention policy measures help create sustainable economies, and as a consequence contribute to human development. There is another component that we sometimes forget. Take the case of vaccines and immunization campaigns as mentioned earlier, biological security issues cannot be taken for granted. The recent epidemic of H1N1 demonstrates that one cannot underestimate even the biosecurity aspect that vaccination campaigns can contribute to. What I care most about, relates to research and innovation in life sciences, including the discovery and development of new medicines and vaccines. Does it contribute to sustainable development? Absolutely! The progress that mankind has made in the last 100 years, thanks to vaccination, and thanks to the availability of new medicines that has had a major impact on demographic evolutions, and on healthier and wealthier societies. In a nutshell, you can recall that there is an extensive literature that can be quoted in this context, revolving around "health equals wealth" (Alsan et al. 2008).

PM: **Reference to the triangular relation between health, education, and wealth, for instance in the works of Jeffrey Sachs.**

Dr. Pisani: Well, besides Jeffrey Sachs[4], there have been a few other academics that have maintained similar positions, and the European Commission has promoted the same concept of health and wealth for quite a few years. Now you can hear it in several other emerging economies, and yet not everybody walks the talk. But it is a very critical concept that has been acknowledged and validated through some empirical research. The economic factors that stem from research, development, and investment in innovation, also have to take into account diffusion of technologies. Diffusion of technologies is another component of sustainable development. In fact, I think the vaccine industry has been one of the first to transfer technologies in different parts of the world in partnership with governments or domestic organizations, in order to ensure that for public health and biosecurity reasons, there could be an appropriate infrastructure and resources for all patients. That is another example that I classify under the economic dimension.[5] Diffusion also means availability and accessibility of health technologies to all patients who need them. A recent study evidenced that the pharmaceutical sector has roughly generated US$ 441 billion worldwide in direct gross added value,

---

[4] http://jeffsachs.org/category/topics/sustainable-development/.
[5] http://www.ifpma.org/fileadmin/content/Global%20Health/Vaccines/Elsevier-Delivering_the_promise_of_the_Decade_of_Vaccines.pdf.

equivalent of the economic strength of Argentina alone for 2011. Another value no less important is that this industry employs some 4.2 million people worldwide, the equivalent of Austria's labor force.[6] There is no doubt that innovation is vital for the economic development of a given society through a productive workforce, through the ability of children not only to survive scourges of diseases but also to grow and learn making their way into the school system and the ability to later on contribute to the community, to be able to undertake economic activity and attract investment. Again, if we go back to vaccines, they do more than just protect health; they also protect incomes and promote strong economies through direct and indirect cost savings. In addition, vaccination has been highlighted as one of the main reasons for the fall in health disparities both within and across countries in the last century. ***In a nutshell, innovation is not only saving people's lives and making them healthier but also generating high-value opportunities the world needs.*** This brings me to my third point. You know that the environmental dimension is a critical one. How to manage appropriately the resources of the planet and the managing of it ranges from appropriate manufacturing practices to appropriate use of energy and protection of the environment in all its components, from water to air; how to build smarter technology systems, convert current environmentally unfriendly options with renewable and clean energy in the most effective ways to reduce polluting emissions, make advances in better urban designs, use innovative modes of transport, etc. In our field, we have to meet stringent regulatory demands to ensure that our business is delivering state-of-art, effective and quality products, all run according to certain rules and procedures.

PM: **Reference to the nature of manufacturing biologicals such as vaccines and the consequences in terms of quality controls and protection of the environment.**

Dr. Pisani: As you can imagine, the more we move towards biological medicines, the more this kind of containment is necessary around the plants and laboratories. It is part of our good manufacturing practice and our ethical practices to ensure that environmental protection stands out as one of those societal goals that we contribute to as an industry. Again for vaccines, the journey is complex and it comes at a cost.[7]

PM: **Question on the pace of change in the industry in light of the evolutions of the regulatory processes at FDA and EMA levels, aim-**

---

[6] http://www.ifpma.org/fileadmin/content/Publication/2014/wifor_feasibility_study_2013.pdf.

[7] http://www.ifpma.org/fileadmin/content/Publication/2014/IFPMA_Complex_Journey_Vaccine_Infographic_2014.pdf.

ing at accelerating access through specific regulatory pathways, considering that the industry usually errs on the side of caution as innovative products are rare and treasured.

Dr. Pisani: I would always like to put things in context, and in this case in the context of sustainable development. *Whether we are talking about regulatory issues, policy measures, or public health at large, there is one thing that is absolutely imperative—constant dialogue between all stakeholders and concerned parties.* Without a continuous dialogue and opportunity of confrontation between government, agencies, civil society, and the industry, there will not be a shared direction of a coherent development. We risk in many instances to be stuck in our own rhetoric. This applies to government, NGOs, and the pharmaceutical industry. The opportunity to evolve, particularly a regulatory pathway, has to take into account a number of product development data that relate essentially to chemistry, manufacturing and control of products, and preparation of regulatory submissions. Usually, the innovative pharmaceutical industry is not lagging behind when it comes to technical and regulatory matters. Our companies contribute to some of the pilot phases to gauge these regulatory pathways' feasibility and efficiency. This collaboration is a continuous process that is prolonged through the implementation phase, where consensus industry feedback on real-time use of these pathways by a larger number of companies and products is shared with regulatory authorities for potential adjustments or improvements. Long-standing industry engagement with regulatory authorities aims at ensuring that there is a common platform with regulators in order to secure that steps are taken to speed up regulatory processes and to make those as efficient and effective as possible. All stakeholders have, as a common goal, to ensure new therapies are available to patients in the shortest possible time frame. So, if sustainable development also means availability or accessibility of new treatments to patients around the world, then the regulators also have a responsibility to ensure that a consistent, coherent, and efficient regulatory framework is in place to achieve this goal.

PM: **You are raising a key point, that dialogue should be continuous. Would you suggest that the industry should adjust some of its processes to manage these interactions and their content?**

Dr. Pisani: Today, there already exist opportunities of interaction and exchanges also on a purely advisory basis between regulators and manufacturers. Informal and formal interactions, as appropriate, should be encouraged and made available to manufacturers. Besides the major pharmaceutical companies, there are hundreds of medium-sized companies, which do not have the same level of resources, whose existence heavily depends on a constructive interaction with regulators. But not only. The industry must also count on the *media* to help dispel misconceptions about the pharmaceutical industry and provide an accurate account of the contributions of pharmaceutical research and development to human health and well-being. The media must therefore play its part in critically examining public perceptions and industry realities. The media can thus help by providing a fact-based account of how the pharmaceutical industry works and the challenges it faces. The public in general long to learn more for instance on topics like why pharmaceutical R&D productivity has declined, i.e., spending too much on too little output; where pharmaceutical companies need to invest their resources; what can be done to solve core health challenges, including cancer, diabetes, and neurodegenerative diseases; and last but not least how the pharmaceutical industry can regain public trust and improve its image. By spreading the word on all these points may prompt a genuine change in the negative perception we usually face.

PM: **It is all the more critical as growth and innovation is increasingly coming from these small companies, which are not making the same organizational choices as larger companies, either by choice or by necessity.**

Dr. Pisani: There has to be an understanding that, if we want to favor an ecosystem which is thriving and which fosters biopharmaceutical innovation, all actors have to obtain the same guidance and advice from regulatory authorities. Perhaps it is not about changing processes in a dramatic way, it is always about changing culture and behaviors in a way that allows the appropriate platforms to be established at multiple levels. In reference to the multilateral public health dimension, for instance epidemiological research, the fact of allowing more regular public–private collaboration, partnership on specific issues and projects of public health benefit, would be fundamental and that is not always supported or understood by all relevant parties.

PM: **Point on the need to change the behaviors and mindsets, not just the processes.**

Dr. Pisani: When thinking about the goal of collaboration and partnership, what are the key challenges? One is in relation to trust building. There should be a culture and behaviors that are conducive to an open dialogue and potential for partnership and collaboration. Other hurdles may derive from governance, as regard to processes. Governance should not be static and be an obstacle to achieving certain goals. And last but not least, availability of resources has to be also clearly indicated. Industry has to put appropriate resources to guarantee quality, safety, and efficacy of medicines. Regulators have to put in place the appropriate infrastructure, including human resources, to ensure that regulatory reviews are timely and professional. So, it depends on the angle that you are looking at, but I will not lose sight of the ultimate concept that you are exploring, which is about sustainable development, and intended in this case, as healthier societies in the world. This is the concept that, in my opinion, must appear in the future UN development goals. Achieve healthier societies, and all the elements we have just talked about. There are numerous examples of best practice that can help select the appropriate tools to generate positive outcomes in fields such as infectious diseases, and notably neglected tropical diseases. Let us think about what is happening, thanks to mobile health[8]. If you think about prevention activities and programs, mHealth tools for example allow focusing on health risk factors for chronic non-communicable diseases. On all these fronts, we are well engaged both as an industry association and as individual companies. There are some 240 partnership programs[9] that have an impact on sustainable development and healthier societies. There are over 50 programs only in the year 2013 for mobile health applications relating to non-communicable diseases. This is a concrete example of how our industry is implementing concrete novel solutions to address broad challenges.

PM: **Thanks for the time and contribution.**

References

Alsan M, Bloom D, Canning D, Jamison D. The consequences of population health for economic performance, in health, economic development and household poverty, from understanding to action. 2008. p. 21–39.

---

[8] http://www.ifpma.org/news/news-releases/news-details/article/ifpma-joins-international-telecommunication-union.html.

[9] http://www.ifpma.org/fileadmin/content/Publication/2013/IFPMA_Health_at_your_fingertips_July2013.pdf.

Chapter 10
Sustainable Development Initiatives: Examples of Successful Programs and Lessons Learned— Interview with Dr. François Bompart

Pierre A. Morgon

The discussion with François Bompart provided a detailed, pragmatic, and action-oriented account of the tools that are part of the corporate responsibility programs, such as tiered pricing in the specific context of a malaria-focused initiative, and explored the recent evolutions since the early programs which were derived from human immunodeficiency virus (HIV) politics and market dynamics, and their impact in terms of corporate governance.

Pierre Morgon: What's your opinion about the role and place of sustainable development in the healthcare industry? How does it contribute to its evolution and what it means in your scope of accountability?

Dr Bompart: The question needs to be put in perspective with the specific expectations of the stakeholders of the health-care industry in light of the very specificity of what the industry is working on, which is human health. It thinks the expectations are higher for our industry than for other industries, given the number of philosophical questions related to life, death, ethics, human rights, and to a form of original sin since this industry is generating revenue from dealing with human ailments.

This research-based industry's business model is today extremely challenged first by the increasing focus on transparency of prices, data, etc., and second by the movement towards Universal Health Coverage. Universal Health Coverage is certainly a goal that all human beings rightly aspire to, but it raises issues for the industry that need to be addressed. The key question is "how

---

P. A. Morgon (✉)
Theradiag, Croissy-Beaubourg, France
e-mail: pierre.morgon@wanadoo.fr; mrgn@bluewin.ch

Eurocine Vaccines, Solna, Sweden

AJ Biologics, Kuala Lumpur, Malaysia

do you ensure universal access to care and medications, while maintaining profitability so as to keep funding R&D?"

I will take a perspective related to corporate social responsibility (CSR) linked to the four prevailing reasons for CSR explained by Michael Porter in his article published in 2006, namely the moral obligation, sustainability, license to operate, and reputation. To those four imperatives, I will add a fifth dimension, which is innovation.

Let me take the example of malaria research. In the case of this particular disease, there isn't a strong incentive for life science companies to generate significant revenues and that allows for experimentation on innovative ways to search for health-care solutions. The treatments are now very cheap—a handful of industry players are investing in diagnosis, training of healthcare providers, strengthening primary care infrastructure, but if profitability is not maintained at adequate levels, the entire model is put in danger by the growing concern expressed by some countries that they do not want to pay more than their neighbors, even though the latter may be poorer than the former. Tiered pricing has been largely practiced, in particular in the field of vaccines, but implementation of intra- or intercountry differential pricing is a challenge as it is not deemed acceptable at similar levels by all the stakeholders. This approach was valued while the rich countries were wealthy to the point that they were not concerned by bankrolling health-care procurement in poorer countries. As their health-care system's financial circumstances are increasingly challenged, even rich countries may become reluctant to foot the bill of funding innovation as they used to, even in the case of rare diseases. As a consequence, several stakeholders are asking the industry to explore ways to de-link the cost of R&D, and the price of its medications.

One needs to look at history to understand the origin of those questions and of the current attitudes and beliefs. From the vantage point of its driving principles, sustainable development has been formatted by the HIV politics and market dynamics. One will recall that, in the late 1990s, a strong movement emerged to tackle this disease, overwhelmingly prevalent in emerging countries, and that the high price of antiretrovirals resulted in a fierce argument over compulsory licenses, technology transfer, and intellectual property rights, all of those linked to the price of medications to the countries and the end users. The arguments opened up a new era in which research-based pharmaceutical firms understood that it was in their best interest to design initiatives to enable manufacturers based in emerging countries to manufacture cheaper alternatives to the original medications.

Nowadays, CSR initiatives have become a way for the industry to retain a seat in ongoing works and exploration, and most importantly to have the legitimate right to experiment new approaches aiming at improving access to care. This is critically important if the industry wants to be part of the discussions pertaining to Universal Health Coverage, as the solidarity model which is the foundation of the ongoing thinking, must also include mechanisms to sustain funding for R&D and for innovation.

PM: **Could you describe succinctly the sustainable development initiatives you've been involved into? What were the specific hurdles to overcome and practices to apply? How eventually does it bring value? What do you think would be important to implement them successfully?**

Let us explore the options available to practice tiered pricing. Although this is a concept that has been successfully implemented in specific markets, such as vaccines, it is difficult to implement on a large and sustainable basis. As a matter of fact it is raising a host of issues, such as the implementation of regulations on free circulation of goods when price differentials may elicit parallel trading in conditions that are not always satisfactorily controlled, complex discussions over reference pricing, transparency issues, etc.

Going back to my earlier example on malaria, the major result of the Sanofi initiative, in partnership with the Drugs for Neglected Diseases initiative (DNDi), has been the ability to procure courses of therapy for about US$1 for an adult and half a dollar for a child, hence a dramatic improvement compared to previous treatments which were three to four times more expensive. In addition, Sanofi has contributed tools and expertise for health-care practitioners' training as well as education campaigns to help all the players in the field, ranging from nurses in bush outposts to children in schools. The package that is offered by Sanofi is not only comprehensive but also "a la carte" since the countries can pick and choose which elements of the program they are interested to implement. This is a kind of activity that only a large health-care company with the relevant medical expertise can execute successfully. The first learning point of that experience was that it was of paramount importance to work with DNDi, which to a large extent can be considered as a challenger of the industry. This enabled Sanofi to aim for a target price for its antimalarial medicine much lower than it would have dared set for itself, and to see the value of not seeking patents in this specific situation (Bompart et al. 2011; Pécoul et al. 2008).

The second learning point was that the system functioned well since the malaria market is heavily subsidized; hence, market dynamics are rather predictable. The risk is that when the subsidies dwindle, the system may run out of steam. Even though the Sanofi–DNDi initiative set the lowest possible prices for its antimalarial medicines at the time of launch, Sanofi is progressively challenged by cheaper alternative manufacturers as price becomes the sole decision factor. This raises significant issues for the continued involvement of research-based companies when faced with low-cost generic competition, since, in addition to providing affordable drugs, multinational players also provide a large variety of added services, such as educational initiatives, sustained R&D and pharmacovigilance programs, etc. Going forward, it will be up to the research-based industry to ensure that the discussions with the country-level stakeholders revolve around what is the best deal for everyone, as opposed to solely focusing on what is the cheapest price for commodities.

The "benefits" that Sanofi derived from this initiative, beyond the obvious visibility, include an increased access to key stakeholders not only in emerging countries but also and most importantly in donor countries and organizations. It also demonstrated the benefits that can be derived from working in partnership with external partners such as nongovernmental organizations (NGOs), governments, funding agencies, etc.

Yet, it is critical to remember that those working relationships are fragile and that it takes very little to shatter them to their foundations: A single issue that is blown out of proportion by the buzz in the media could damage the working spirit almost beyond repair.

Without any hint of provocation, it is important to bear in mind that there is always an element of public posturing; and as a consequence, the attitudes of the partners are oftentimes different between that within the working groups and the public face in front of the media. All too often, the industry is accused of duplicity by its challengers who often structured their communication around simple and abrasive messages so as to catch attention of the media, public opinion, and political decision makers.

PM: **What would you recommend for the approach to be sustainable?**

Dr Bompart: In my opinion, the best solution for the research-based industry is to keep doing what it does best: Keep bringing innovation to the table. For instance, this could mean exploring other diseases than those currently under the limelight. Sanofi is currently developing access programs for patients with central nervous system (CNS) disorders such as epilepsy, depression, or schizophrenia in developing countries, and is exploring the possibility to set up tiered pricing across countries. Such disease are chronic, oftentimes lifelong, and they will require innovative models to be sustainable in the long term.

They lend themselves well to such an exploration since most of the treatments are already off patent.

The main obstacle in looking at those diseases is to get to the right level of awareness, attention, and commitment of political forces which tend to be ranking them quite low in their list of priorities.

These diseases are also interesting because they will drive all the stakeholders to go beyond the systems of "vertical funds" that were created for transmissible diseases, such as acquired immune deficiency syndrome (AIDS), malaria, or tuberculosis, to reach the point of discussing the strengthening of the health-care systems that will be required to successfully manage all diseases, including those chronic diseases that are nontransmissible.

The major life science corporations have all the expertise it takes to bring a lot of value in those emerging projects. Should they want to remain a major partner at the discussion table, they will not only need to participate through innovation and added services but also through superior quality of its medications and its supply chain. Patients in resource-poor countries, as well as public authorities, are especially exposed to the risks posed by products of substandard quality and counterfeit medicines.

Discussions are currently taking place at the global and regional levels aiming at harmonizing quality standards, ranging from good manufacturing practices (GMP) to registration and pharmacovigilance in many parts of the world, such as those revolving around the Pharmaceutical Manufacturing Plan for Africa (African Union Pharmaceutical Manufacturing Plan for Africa 2012).

These initiatives are of utmost importance for the research-based industry. Indeed, they have the potential to either deepen differences in standard requirements between regions or create a global level-playing field, ensuring fair competition based on similar regulatory standards for all manufacturers. This latter option is the one that would best benefit patients, countries, and funders. It requires the active involvement of all stakeholders, including that of industry as a key and responsible partner for global health.

References

African Union Pharmaceutical Manufacturing Plan for Africa: African Union Pharmaceutical Manufacturing Plan for Africa. 2012. http://apps.who.int/medicinedocs/documents/s20186en/s20186en.pdf. Accessed: 14. Apr. 2014.

Bompart F, Kiechel JR, Sebbag R, Pécoul B. Innovative public-private partnerships to maximize the delivery of anti-malarial medicines: lessons learned from the ASAQ Winthrop experience. Malar J. 2011;10:143.

Pécoul BS, Amuasi AM, Diap J, Kiechel GJR. The story of ASAQ: the first antimalarial product development partnership success. In Health partnerships review: focusing collaborative efforts on research and innovation for the health of the poor. 2008. pp. 77–83

Chapter 11
The Challenges of Sustainable Development for the Health-Care Industry: An Examination from the Perspectives of Biomedical Enterprises

Geoffrey Chun Chen

Introduction

When biotechnology began to play a role in industrialization, it started from the food and beverage industry, such as alcoholic fermentation. With the rise of life sciences in the late twentieth century, we keep redefining the boundary of biotechnology industry. Biotechnology has enabled the revolution of a wide range of industries such as health care, agriculture, and energy. Bioenergy and bioremediation could contribute to the overall human health by controlling disease-causing agents, while bio-agriculture can improve global health by reducing famine. These can indirectly enhance the sustainability for the health-care industry. Nevertheless, we pay greater attention to biomedical products and services that have stronger ties with the health-care industry.

Sustainable development in the health-care industry is aimed at ensuring the basic rights of human beings over both the prevention and treatment for all kinds of diseases for the future generations to come, which are typically defined as seven generations.[1] The basic needs include the access to the standard of care for every human being in an efficient and affordable manner. Typically, we want to understand whether the current practices and the future trends can provide sustainable health-care services to the ever-growing population. We also want to make sure our policies, either from the private sector or from the government, that will have long-run strategic impacts instead of just meeting short-term needs. In the theory of sustainable development, the classical three-pillar concepts, "economic," "social," and "environmental," have profound applications in the health-care industry. Therefore, we would like to analyze how biomedical enterprises could address these challenges for the health-care industry and how biotechnology gives rise to some new problems, such as bioethics and regulation over new technologies.

---

[1] Constitution of the Iroquois Nations: The Great Binding Law, Gayanashagowa. (n.d.).

G. C. Chen (✉)
BioMaryland, 3501 Saint Paul street, apt. 423, Baltimore, MD 21218, USA
e-mail: geoffrey.chen@jhu.edu

© Springer International Publishing Switzerland 2015
P. A. Morgon (ed.), *Sustainable Development for the Healthcare Industry*,
Perspectives on Sustainable Growth, DOI 10.1007/978-3-319-12526-8_11

First, from the economy perspective, we argue that the prosperity of the healthcare industry is being challenged as it takes a longer time and higher cost to bring new solutions to the patients. We might wonder whether there will be a turning point of this trend and when it would come.

During the golden period of the pharmaceutical industry until the late 1990s, the large companies bore most of the risks in innovation and thus enjoyed high returns protected by the patent law. As the output of research and development (R&D) is dropping, the innovation of pharmaceutical companies nowadays relies on developing strategic partnerships, constructing licensing deals, or pushing merger and acquisition with biomedical firms that are expected to have some disruptive products for the unmet medical needs. The biomedical innovation will be the main contribution to the next boom of new therapies. As a whole, the biotechnology community carries greater responsibility of innovation for the whole health-care industry. In addition, the biomedical industry has extended the areas of research such as bioinformatics, epigenetics, tissue engineering, stem cell therapy, gene therapy, and so on. In a sense, we are just embarking on the way to decode the secret of life. Biomedical discovery will bring the promising future of the health-care industry. However, what makes the sustainable development challenging for biotechnology companies is that it requires a continuous and tremendous amount of funding resources for a very long duration. In most of the countries, governments or universities typically administer and allocate the budget of basic research. From the perspective of start-ups, the risk is high all along the way from preclinical research to the launching of the products in the market. Investing in a wide range of start-ups is the way in which venture capitalists can hedge their bets and manage their risk of their portfolio. However, when the global economy derails from the robust growth and stability, the venture investments in biomedical companies require more patience and encouragement.

Second, from the social perspective, the biomedical industry development can continuously improve the health status of mankind. Better health will ensure one's higher productivity and increase social welfare as a whole. Obviously, the standard of care is always progressing as the new biomedical treatments become available. As we expect to live longer, we naturally care more about the future. Sustainability is a rewarding area for people to pursue. For example, we have seen many corporations adopting the concept of "citizenship," in which employees are keen on participating in various initiatives to save resources, increase fairness, help the community, etc. Nevertheless, from the social perspective, one of the great challenges over the sustainability is how to control the disparity of access and control the health-care spending, once a new medical breakthrough is achieved. For example, the genetic sequencing gives rise to some new problems such as discrimination in employment, marriage, and health-care insurance.

Third, from the environmental perspective, the biomedical industry still faces traditional challenges, such as green manufacturing process and medical wastes control, just to name a few. The evolution of viruses is unknown knowledge to us, and this could be regarded as a response of nature to our scientific advancement in vaccines or antibodies, which poses a potential threat to our sustainability in the health-care sector. Even though we might not change the course of virus evolution, we need to reserve more resources to deal with catastrophic events. For example, the HIV has a severe impact on the world's economy. Something that is more scary is the concept of artificial biodiversity. No one knows for sure how we break the balance of nature when we introduce the genetically engineered species to the ecosystem or fail to control them among our research laboratories. However, the solutions to these problems have to be nothing but biotechnology itself.

Sustainable Development Framework for Biotechnology Enterprises

Sustainable Development

According to *Our Common Future,* sustainable development means the development that meets the needs of the present without compromising the ability of future generations to meet their own needs. It contains within it two key concepts: the concept of needs, in particular the essential needs of the world's poor, to which overriding priority should be given and the idea of limitations imposed by the state of technology and social organization on the environment's ability to meet present and future needs (World Commission on Environment and Development (WCED) 1987). All definitions of sustainable development require that we see the world as a system—a system that connects space and a system that connects time.[2] As the global ecosystem, economy, and value system become a more coherent piece, sustainable development will require more international collaboration. Nowadays, information technology has enabled easy sharing and documentation of our sustainability achievement; we cannot let our future generations down.

Many public companies have understood well the importance of the sustainable development. They created their framework of sustainable development not only to address the expectation of investors or employees but also to align with their business interests. The theory behind this is the sustainability of the sweet spot: the place where the pursuit of profit blends seamlessly with the pursuit of the common good. The best-run companies around the world are trying to identify and move into their sweet spots. Moreover, they are developing new ways of doing business in order to get there and stay there (Savitz and Weber 2006).

Fortune 500 companies usually operate at a global scale. Their business is closely related to a large number of consumers. Thus, it is important to develop the sustainable development that promotes the image of the company and nurtures a channel to communicate with prospective customers. For example, Procter & Gamble has developed a series of campaigns around the sustainability issue of water usage.[3] Naturally, Procter & Gamble have the best resources such as technology and talent to meet the social needs while keeping benefits for its long product lines for personal care.

Meanwhile, most companies provide sustainable development reports, in which those numbers of energy conservation and recycled waste are compared each year.

---

[2] What is Sustainable Development? (n.d.). Retrieved June 18, 2014 from http://www.iisd.org/sd/.

[3] P&G's Sustainability Vision includes environmental sustainability and social responsibility. (n.d.). Retrieved Jun 18, 2014 from http://scienceinthebox.com/environmental-sustainability-goals.

Table 11.1 Category and aspects of the guideline in GRI

Category	Economic		Environmental	
Aspects[III]	Economic performance Market presence Indirect economic impacts Procurement practices		Materials Energy Water Biodiversity Emissions Effluents and waste Products and services Compliance Transport Overall Supplier environmental assessment Environmental grievance mechanisms	
Category	Social			
Subcategories	Labor practices and decent work	Human rights	Society	Product responsibility
Aspects[III]	Employment Labor/management relations Occupational health and safety Training and education Diversity and equal opportunity Equal remuneration for women and men Supplier assessment for labor practices Labor practices grievance mechanisms	Investment Nondiscrimination Freedom of association and collective bargaining Child labor Forced or compulsory labor Security practices Indigenous rights Assessment Supplier human rights assessment Human rights grievance mechanisms	Local communities Anticorruption Public policy Anticompetitive behavior Compliance Supplier assessment for impacts on society Grievance mechanisms for impacts on society	Customer health and safety Product and service labeling Marketing communications Customer privacy Compliance

GRI global reporting initiative

They usually set specific goals with a transparent measurement mechanism. Some companies have adopted the Global Reporting Initiative (GRI),[4] which is a comprehensive sustainable reporting framework that provides metrics and methods for measuring and reporting sustainability. See Table 11.1.[5] We discuss in detail about how biotechnology firms are adopting the GRI framework to engage in the

---

[4] GRI G4 GUIDELINES PART 1 REPORTING PRINCIPLES AND STANDARD DISCLOSURES, Retrieved June 21, 2014, from https://www.globalreporting.org/resourcelibrary.

[5] GRI G4 GUIDELINES PART 1 REPORTING PRINCIPLES AND STANDARD DISCLOSURES Retrieved June 21, 2014, from https://www.globalreporting.org/resourcelibrary.

11 The Challenges of Sustainable Development for the Health ...

Table 11.2 2013–2014 industry group leader from DJSI (http://www.sustainability-indices.com/images/130912-djsi-review-2013-en-vdef.pdf)

Industry group leaders (2013–2014)	Industry group
Volkswagen AG	Automobiles and components
Australia & New Zealand Banking Group Ltd	Banks
Siemens AG	Capital goods
Adecco SA	Commercial and professional services
Panasonic Corp	Consumer durables and apparel
Tabcorp Holdings Ltd	Consumer services
Citigroup Inc	Diversified financials
BG Group PLC	Energy
Woolworths Ltd	Food and staples retailing
Nestlé SA	Food, beverage, and tobacco
Abbott Laboratories	Health-care equipment and services
Henkel AG & Co KGaA	Household and personal products
Allianz SE	Insurance
Akzo Nobel NV	Materials
Telenet Group Holding NV	Media
Roche Holding AG	Pharmaceuticals, biotechnology, and life sciences
Stockland	Real estate
Lotte Shopping Co Ltd	Retailing
Taiwan Semiconductor Manufacturing Co Ltd	Semiconductors and semiconductor equipment
SAP AG	Software and services
Alcatel-Lucent SA	Technology hardware and equipment
KT Corp	Telecommunication services
Air France-KLM	Transportation
EDP—Energias de Portugal SA	Utilities

DJSI Dow Jones sustainability index

sustainable development issues.[6] The Dow Jones Sustainability Index (DJSI) typically tracks sustainability performance by geographical area and industry group. See Table 11.2.

The Biotechnology Version of Sustainable Development (500)

We understand that the current classical approach to evaluate the performance sustainability of companies is to pick a one-for-all framework, whether GRI, DJSI, or

---

[6] ABOUT SUSTAINABILITY REPORTING Retrieved June 21, 2014, from https://www.globalreporting.org/information/sustainability-reporting/Pages/default.aspx.

others, and then compare the indicators among all companies or across the industries. However, if we try to use a metaphor here, this approach seems to order the same version of tests to science students, business students, and medical students to understand their academic performance. It might be more meaningful to see a test tailored for their academic features.

Some can argue that large public companies, regardless of industries, would find much common ground in sustainability issues. However, those smaller firms might feel overwhelmed. When we think about the biotechnology industry, we have to agree that it has a long-tail market feature, which means that the sum of small-sized firms contributes to the economy no less than the big firms do. Therefore, it is questionable that the current comprehensive framework of sustainable development is suitable for the entire biomedical industry to adopt. For middle-sized companies that are in the transition stage thinking about the sustainability, if we visit the GRI framework (Table 11.1), we might wonder how many aspects they can align with their current activities. For example, we should understand how to measure their contribution over sustainability, such as identifying the mechanism of the disease, saving health-care expenditure in the long run.

In this chapter, we are too premature to arrive at the conclusion of what exactly the biotechnology version of sustainable development should be. Having a different version for each industry could be a daunting task. However, this industry is so unique that a different alternative should be explored. In the next section, we examine companies in different scales over what they have accomplished and how they have promoted over the sustainability issues.

Industry Initiatives

Biomedical Giants (400)

Genentech & Roche Genentech has a great emphasis on environmental sustainability. Mainly, the company provides sustainability commitments in efficient energy use, water conservation, and waste reduction. The company priority on sustainability depends on its events and its ecological environment. For example, Genentech is located in Southern California, where water is scarce. Therefore, the company is paying extra attention to the water usage. In terms of energy reduction, between 2009 and 2012, Genentech has seen an increased emission from air travel. That was due to increased international travel after its merger with Roche. More alternatives, such as using virtual meeting technology, have been adopted.

Patient access is also a highlight in Genentech's sustainability. Since Genentech's first product launch, the company has provided 3.5 billion in free medicine to uninsured patients. The Genentech access to Care Foundation and Genentech Therapy-Specific Co-pay Cards are patient assistance programs through which the business interests could be addressed, simultaneously. Overall, Genentech communicates its sustainability issues for the sake of patients and employees.[7]

Roche has a different way to express its sustainability. It believes that only environmentally and socially responsible companies can achieve sustainable financial success. It claimed their daily work in the R&D to be their most significant contribution to society. Roche also mentioned the importance of being transparent with regulators, customers, and suppliers. In 2013, the DJSI as the group leader within the pharmaceuticals, biotechnology, and life science industry recognized Roche. Unlike Genentech, Roche employed the GRI, reaching A+ as the best result.[8]

Abbott

Abbott is one of the most diverse, global health-care companies. The sustainability issues were raised in its global citizenship report.[9] The essential idea "Turning Science into Caring" mentioned that Abbott's strategies for business growth and profitability as inseparable from its strategies for citizenship and sustainability. The priority is R&D. Abbott believed that solving global health-care challenges with the sustainable solutions is a part of sustainability. The second priority is to provide promising products. The third one is to ensure patient access. The last part of the emphasis is "safeguard the environment." Interestingly, Abbott is not only taking care of its own water usage but also actively reaching out to the community. For example, the Abbott Fund has expanded their partnership with Project Water Education for Teachers (WET) to educate children about saving water. What is special is Abbott's sustainable packaging, including an increase in reuse, sourcing packages with more renewable energies. In brief, Abbott is one of the very few biopharmaceutical companies with a very detailed strategy framework to address sustainability issues.

Biogen Idec Inc.

Biogen Idec Inc. ranked second in the world's 100 most sustainable companies (see table below). Its sustainable strategy is embedded in the corporate citizenship. Company image in the minds of employees is very important for Biogen. For example, Biogen has the Sustainability Leadership Award that honors employees around the world who have implemented projects aimed at reducing our environmental impact. In addition, Biogen has Biogen Idec Foundation to support medical and science, technology, engineering, and math (STEM) education, provide humanitarian assistance, and fund important community projects.[10] Biogen also has a greater consideration of environmental issues. For example, its Building 26 applied LEED certification, which is a voluntary program established by the US Green

---

[7] Genentech: Good. (n.d.). Retrieved June 20, 2014, from http://www.gene.com/good.

[8] Roche – Reporting and Indices. (n.d.). Retrieved June 20, 2014, from http://www.roche.com/responsibility/sustainability/reporting_and_indices.htm.

[9] Abbott Global Citizenship Full Report. (n.d.). Retrieved June 20, from www.abbott.com/static/cms_workspace/content/document/Citizenship/2011/Abbott_Global_Citizenship_FullReport.

[10] Improving Lives. (n.d.). Retrieved June 20, 2014, from http://www.biogenidec.com/improving_lives.aspx?ID=14606.

Building Council.[11] Lastly, Biogen has a special area of sustainability that other companies do not implement: the diverse supplier. Biogen Idec defines a diverse supplier as a business that is women-owned, veteran-owned, lesbian, gay, bisexual, and transgender (LGBT)-owned, service disabled veteran-owned, minority-owned, historically underutilized business, and small business vendors as defined by the US Small Business Administration.[12] Sustainability always creates implicit values hard to measure, and that is why not all companies are willing to commit to it to as they do to the R&D. However, the difference in such devotion provides the opportunity for Biogen to position the company image in that strength. In the company website, Biogen states that the long-term success requires an inspired approach in engaging with stakeholders, advocating for sensible public policy, entering new markets, managing responsibly, and navigating the changing health-care landscape.[13] We believe Biogen has bet on the return on investment in the very long run for its consistent corporate strategy on sustainability.

The world's 100 most sustainable companies, 2014[14]

Rank	Company name	Headquarters	GICS sector	Overall score (%)
1	Westpac Banking Corporation	Australia	Financials	76.5
2	Biogen Idec Inc.	USA	Health care	75.3
3	Outotec OYJ	Finland	Industrials	74.2
4	Statoil ASA	Norway	Energy	74.0
5	Dassault Systemes SA	France	Information technology	74.0
6	Neste Oil OYJ	Finland	Energy	69.2
7	Novo Nordisk A/S	Denmark	Health care	68.8
8	Adidas AG	Germany	Consumer discretionary	68.0
9	Umicore SA	Belgium	Materials	67.8
10	Schneider Electric SA	France	Industrials	66.5

GICS global industry classification standard

---

[11] Rethinking resources. (n.d.). Retrieved June 20, 2014, from http://www.biogenidec.com/rethinking_resources.aspx?ID=11581.

[12] Supplier diversity. (n.d.). Retrieved June 20, 2014, from http://www.biogenidec.com/supplier_diversity.aspx?ID=19372.

[13] Creating value. (n.d.). Retrieved June 20, 2014, from http://www.biogenidec.com/creating_value.aspx?ID=11614.

[14] The World's Most Sustainable Companies Of 2014– Forbes. (n.d.). Retrieved June 28, from http://www.forbes.com/sites/jacquelynsmith/2014/01/22/the-worlds-most-sustainable-companies-of−2014/.

Biomedical Start-ups

The majority of biotechnology start-ups conduct early-stage R&D. Typically, large pharmaceutical companies would acquire some of the successful biotechnology start-ups or simply saying survivors among the start-ups. The lifecycle of start-ups usually lasts less than 10 years and succeeds in either exiting, such as being acquired by big companies or going public, or being dismissed with failure. This nature gives fewer incentives for biotechnology start-ups to focus on sustainability or social responsibility. In terms of sustainability, we seldom see any activities for environmental protection or energy reservation.

Adhezion Biomedical[15]
Adhezion Biomedical is a company in North Carolina that develops and produces cyanoacrylate-based medical adhesive, wound care, and microbial barrier products for connective tissue. It was formed in 2001, and received the first approval of 510(k) in 2007. There is no information about sustainability or corporate social responsibility on their website.

Lysosomal Therapeutics Inc.[16]
Lysosomal Therapeutics is dedicated to innovative small-molecule R&D in the field of neurodegeneration, yielding new treatment options for patients with severe neurological diseases. There is no information about sustainability or corporate social responsibility on their website.

Bicycle Therapeutics[17]
Bicycle Therapeutics was founded in mid-2009 as a spinout of the Medical Research Council Laboratory of Molecular Biology (Cambridge, UK). There is no information about sustainability or corporate social responsibility on their website.

Bicycle Therapeutics[18]
DKIS LLC is a Russian biopharmaceutical start-up company that develops novel hepatitis C virus-like particles. There is no information about sustainability or corporate social responsibility on their website.

Medium-Sized Biomedical Companies

The medium-sized biomedical companies have aroused our interests in particular because most of them would experience a transition period of adopting corporate sustainability strategy. Other scenarios, such as merger and acquisition, first public offering, or spinning out from a large company, could also play a role in shaping the

---

[15] Retrieved June 28, from http://www.adhezion.com/docs/company/default.aspx.
[16] Retrieved June 28, from http://www.lysosomaltx.com.
[17] Retrieved July 8, from http://www.bicycletherapeutics.com.
[18] Retrieved July 8, from http://www.dkis.ru/eng/cont-eng.html.

sustainability issues. Interestingly, we typically find that the companies in the scale have quasi-sustainability strategies, which probably have a more direct relationship with their core business. Subsequently, we have a review of several medium-sized biomedical companies of different kinds to evaluate how they are doing over the sustainability issues.

Avanir
Avanir is a biopharmaceutical company in California focused on bringing innovative medicines to patients with central nervous system disorders. It generated around US$ 75 million in revenue and hired about 300 employees in 2013.[19] We can tell that the company is in the amateur stage of sustainability strategy. No independent "sustainability" section appears on their website. In addition, the company has more emphasis on corporate government than environmental sustainability. Unfortunately, none of the indicators is evaluated in numbers.

Intercept
Intercept is a biopharmaceutical company that focused on the development and commercialization of novel therapeutics to treat chronic liver and intestinal diseases.[20] Since all its products are in the pipeline, they have no revenue yet. The only piece related to sustainability is the corporate governance, which covers the basic aspects such as auditing, nomination, and compensation.

Spectrum
Spectrum is a biotechnology company with fully integrated commercial and drug development operations, with a primary focus in oncology and hematology. It had about US$ 155 million in 2013 with four products in the market. The sustainability is missing for this company. However, corporate governance is the only part that the company emphasized.

23 andme
23 andMe, Inc. is privately held company dedicated to helping individuals understand their own genetic information using recent advances in DNA analysis technologies and web-based interactive tools.[21] The company has been a pioneer in this category, and it is believed that its business can significantly change how we understand our genetic vulnerability to diseases. Currently, FDA is challenging the business of genetic testing targeting consumers. In terms of sustainability, 23andme has almost zero initiatives about corporate governance or social responsibility, let alone sustainability. The company has put majority of their efforts in sales and marketing.

StemCells, Inc.
As the leader in this category, StemCells uses stem cell biology to discover, develop, and commercialize breakthrough therapeutics and enabling tools and technologies for use in stem-cell-based research and drug discovery. The only part related to

---

[19] Retrieved July 8, from http://www.avanir.com/about.
[20] Retrieved July 8, from http://www.interceptpharma.com/about/.
[21] Retrieved July 8, from https://www.23andme.com.

sustainability is in the corporate governance part. However, we all expect that stem cell therapy can significantly increase the sustainability of mankind given its cost-effectiveness and possibility of cures. The CEO and President Martin McGlynn said, "Success in harnessing the full therapeutic potential of stem cells would allow us to address the root cause of the underlying disease rather than just continuing to treat symptoms. The prize would be a paradigm shift that could fundamentally transform the practice of medicine and health-care economics."[22] Nevertheless, stem cells created a lot of controversies in the social aspect.

Ethicon

Ethicon has been part of Johnson & Johnson since 1949, but in 1992, it became a separate corporate entity. In their website, it says, "Throughout our history, Ethicon has remained committed to the Johnson & Johnson goals of improving the health and well-being of the world community." It also uses the tool Earthwards®, a Johnson & Johnson approach, for creating sustainable solutions across the product lifecycle.[23] We believe that the connection to Johnson & Johnson has been a great help to the level of sustainability strategy that Ethicon is playing.

Although the examples above represent only a part of medium-sized biomedical companies, we can say that the sustainability practices have been largely limited to the focus on corporate governance among these companies. Chances are good that some companies that just left the start-up stage have other business challenges to worry. Still, the opportunities exist as some of these companies may start to scale-up in manufacturing. Therefore, environment issues, such as energy reservation, could help with cost controlling. Other aspects include attracting talents in this industry needed to build up a company image with more responsibility.

Discussion

The research on biotechnology companies about their sustainability is limited by whether companies provide adequate public reports. There are very few literatures about comparing sustainability strategy between small, medium, and large companies. Only those biopharmaceutical companies that are well established have more detailed information open to the public. However, if we are able to provide a quantitative survey among these companies, we can understand the status quo of these companies on sustainability.

This chapter here only provides a qualitative assessment other than digging into the data, and has not taken into account geographic differences. We lean on our analysis in developed countries, where the sustainability issues are better articulated among the public information.

---

[22] Retrieved July 8, from http://www.stemcellsinc.com/about-us/business-strategy.htm.

[23] Retrieved July 8, from http://www.ethicon.com/corporate/our-commitment/sustainability/our-products.

Another area to look at could be the historic change in sustainable development within the same companies. This research could be done through tracking the records of companies' initiatives on sustainability over time and interviewing senior leaders who manage sustainability issues in the companies. Thirdly, we can also explore when companies establish their human resources on sustainability, such as hiring a chief sustainability officer.

Conclusion

Multinational companies, such as Biogen, Abbott, or Genentech, which have customers, public investors, or employees as their key stakeholders, care more about their sustainability issues, especially those related to the community they serve. Some small companies might have initiatives of environmental protection, but the company might be too small to have human resources to collect the information and manage the ensuing processes, and the total utilities costs could be so low that the company directly ignores it. However, it is understandable that financial sustainability is a more important consideration for biotechnology companies than other sustainability activities. From the economic perspective, small biotechnology companies are contributing a greater and greater part of innovation to the whole industry. Without these small companies that bear very high risks of success, the large companies cannot ensure their R&D efficiency and innovation sustainability in the long run. The hybrid between big companies and small ones is the most critical part of the sustainability strategy adoption. These medium-sized companies need to make decisions on how to engage with stakeholders as their business grows, specifically whether they should initiate any sustainability strategies, and if yes, how they could seek a balance between internal issues, such as corporate governance and external issues, such as ecological footprints. One thing these companies often forget is how their core business has been supporting the sustainability of our society. We also figure that the framework of sustainability auditing and reporting is not suitable for companies at this stage because its comprehensiveness is overwhelming. Finally, we suggest that industry experts and sustainability subject experts work closely to design some practical assessment tools on sustainability issues, which are customized for the biotechnology industry with health care as the main focus. By restructuring what these medium-sized companies have contributed to, yet fail to market, sustainability issues, companies can easily understand their economic incentives to carry on more sustainable business models.

References

World Commission on Environment and Development (WCED). Our common future. Oxford: Oxford University Press; 1987. p. 43.

Savitz AW, Weber K. The triple bottom line: How today's best-run companies are achieving economic, social, and environmental success-and how you can too. San Francisco: Jossey-Bass; 2006.

Chapter 12
Corporation's Social Responsibility: From the Awareness of Philanthropy to the Demand of Implementation

The Case of Expanscience, Pioneer of a New Generation of Health Care Corporations

Vanessa Logerais

Much more than just a concept, sustainable development has become indicative of a company's ability to create added value for everyone by reconciling economic development and environmental protection for both humans and nature.

For some industry sectors, corporate social responsibility (CSR) is an operational and productive tool of management for nonfinancial performance, and will even undermine the company's freedom and legitimacy of professional exercise, against the backdrop of strengthened European and international regulations and legislations, as well as greater social pressure.

The mission of health care is to provide cures and improve well-being and as such, it affects people's lives in every way. What is more, the pharmaceutical and dermocosmetics industry, exploiting natural resources on which it largely depends economically, is without a doubt, the most confronted with the dual necessity of being exemplary: on the one hand, identify the most probable challenges in the whole value chain with the largest environmental and social significance; on the other hand, execute an effective plan focusing on their most important ramifications.

Within the past 10 years, evaluation criteria and references of the implementation and reporting of the kind of approaches such as Global Reporting Initiative (GRI) or International Organization for Standardization (ISO) 26000 have always listed the so-called materiality as a top requirement. A company is thus evaluated beyond its awareness of social responsibility in its economic and political development model, but also on the already achieved measurable results, the degree of exhaustiveness and consistency of implementation throughout the activity chain, as well as the ability to report to all stakeholders this particularly approach. The complexity lies in the scale of the sector's responsibility to protect our common heritage, in the conceptual, technological, and operational transformation of this responsibility in

V. Logerais (✉)
PARANGONE, 5 bis rue de la Ferme 92100 Boulogne Billancourt, Paris, France
e-mail: vanessa.logerais@parangone.fr

industrial, ethical, and social dimensions, as well as in the necessity of profoundly adapting professional practices accordingly.

The challenge is so huge that few organizations today are known for the credibility and exemplarity of their initiatives. Established in 1950, as an independent French family-owned laboratory boasting a renowned expertise in the skincare and in the treatment of osteoarthritis, Expanscience is the first pharmaceutical and dermocosmetics laboratory to have obtained at the end of 2013 the level "exemplary" according to the Association Française de l'Assurance Qualité (AFAQ) 26000 assessment method, an Association Française de Normalisation (AFNOR) Certification method based on the norm ISO 26000 as the only international norm for CSR assessment. The company is recognized and appreciated for authenticity and consistency of its approach and the remarkable awareness of collective interests. This drives us to further explore its operational model.

Health care and environment: an interaction rehabilitated at the heart of the progress of health-care sector.

Alongside this strengthened requirement, new concepts enter the scope of social responsibility.

Ancestral but for a long time abandoned by public institutions on account of its protean feature, the concept of environmental health care, aimed at striking a direct connection between the living environment pollution and the health and well-being of mankind, emerges among all the professionals of this sector. The protection of health consists not only in treating or curing diseases but also in taking into account all the external pathogenic factors that could damage our health and in promoting practices that could alleviate or avoid pathology and improve the quality of our living environment.

This new paradigm gives a health-care actor like Expanscience an unprecedented citizen responsibility to invest more in this prevention subject when it comes to its relations with its clients and economic partners. This paradigm is also a new challenge of exemplarity in order to take in-depth thinking and concrete actions to a more mature level.

What are the insights driving the company's commitment to this social responsibility approach? How does the company transform this approach into tangible and concrete actions? What are the benefits? And beyond all that, how is this pioneer experience contributing to the foundation of a whole new generation of corporations, which integrate nonfinancial performance objectives into their management to the point where the business practice and economic model of this profession are profoundly transformed?

Dialogue with Karen Lemasson, Director of CSR and sustainable development of Laboratoires Expanscience.

Logerais: Expanscience has been engaged in a program of CSR since 2004. How would you take stock of after 10 years?

Lemasson: Expanscience's engagement is above all a voluntary initiative, which relies on the great awareness and strong will of its CEO, Jean-Paul Berthomé. When the French Global Compact was launched, further to the UN Global Compact, which gathers together for the first time in history business and UN organizations, labor market and civil society around these fundamental principles in the areas of human rights, labor and environment, we made a commitment to ourselves that the engagement of a company like Expanscience should not be limited to its industrial scope. It's by the yardstick of these first deliberations of what could be the fundamental principles of Expanscience's international contribution to ethical and environmental questions, that our company became a member of the Global Compact in 2004, and launched its social responsibility policy.

In 2009, when Expanscience signed the responsible communication chart of the Advertisers' Union in France (UDA), consultation intensified between the board and stakeholders of our company in order to define our responsibility scope. These have quickly led to concrete actions. We organized and prioritized these actions into a strategy of social responsibility (CSR), structured throughout our entire activity chain from manufacturing to commercialization. In 2010, we reoriented our policy and drew up a seven engagements Plan for 2015, those engagements being subject to operational roadmaps.

In 2011, we became a member of the Union for Ethical BioTrade (UEBT),[1] which enriched our approach and allowed us to support and reinforce the credibility of our CSR action plan in each of our vegetal supply chains.

Today, CSR is so much more than just a program, it's an independent dimension of our corporate governance.

Logerais: What are the main stakes of sustainable development implemented in the production of a medical or dermocosmetic product?

Lemasson: We have identified four axes of commitments and actions: reduction of social and environmental impact of our products, which starts from the analysis of their life cycle; application of a responsible purchasing policy especially throughout all our vegetal supply chains; improvement of our environmental practices on our own production site; setting up a social responsibility policy for our collaborators.

The cornerstone of the first axis is the notion of "eco-responsibility" of our products for our patients and consumers. As for dermocosmetics activities, the notion can be interpreted as "ecodesign." This seeks to take all the lifecycle steps of our products into consideration. It starts by assuring that our raw material supplies come from "responsible" channels and guaranteeing, by the design-research, a greater naturalness (average of 92% of the natural ingredients in the Mustela Bébé range) as well as formulations that are both effective and safe without harmful ingredients to people and environment.

The next step is production. It deals with the impact of manufacturing processes, and includes the reduction of the weight of the bottles and tubes packaging. Then, the transportation step is where we strive to optimize the volume of products transported and reduce greenhouse gas emissions. The usage step is also taken into consideration because of its significant environmental impact, especially for products requiring to be rinsed.

---

[1] Union for Ethical Biotrade (UEBT) is a nonlucrative international association, promoting ethical sourcing practices of ingredients coming from biodiversity. Becoming a member of UEBT requires an audit led by an independent organization according to seven principles including the respect of biodiversity, human rights, traditional knowledge, business ethics, etc.

We believe that the product can act for our consumers as an awareness enhancer for reducing environmental impact and can spread responsible and eco-friendly consumption behavior. Regarding the step of "product end-of-life," it's also designed from the design phase, aiming at ensuring biodegradability of washing formulas[2] and making packages recyclable, subject to consumer compliance with recycling rules. In 2015, 100% of our new dermocosmetic products will be ecodesigned.[3]

When it comes to our pharmaceutical activity, we have analyzed the lifecycle of our major medicine, particularly, the manufacturing process of avocado oil, which is the source of one of its active ingredients and we are striving to reduce environmental risks and impacts linked to the production of this product.

The second axis focuses on our supplies. It is a very ambitious objective because by the year 2015, Expanscience will have carried out a CSR action plan upon 100% of its plant supply chains. Its roll-out is resting on a referential derived from ISO 26000, from the UEBT criteria, and from the Nagoya protocol[4] dedicated to the genetic resources of the planet, endorsed in 2012.

This axis is particularly valuable for the structuring of Expanscience and includes a large scope of actions, from responsible exploitation of natural products, such as avocado, conservation of the biodiversity, training of local professional populations, all the way to combatting the desertification or biopiracy.

The third axis aims at significantly reducing our consumption of gas, water and electricity, and our output of waste and greenhouse gas to reach our reduction goal of 20% between 2010 and 2015.

Our production site in Epernon, France, is home to Expanscience's global industrial facility, from research and development to packaging of products. In 2013, 58,000,000 units of products were manufactured and it was granted ISO 14001 certification in 2012.[5] Moreover, a new building was constructed in compliance with the high environmental quality (HEQ) approach, which allows us to collect 1000 m^3 of rain water per year and ensure 94% of offices heating by recuperated energy. In

---

[2] OECD 302B method.

[3] In 2013, Expanscience's policy implemented from 2007 led to a reduction in the quantity of used material: 100 tons of paper and carton, 40 tons of unspoiled plastics. 100% of the dermocosmetics cases for Mustela and Noviderm are printed with ink made by plant oil, and 100% of the bottles of our major brands are recyclable (according to French criteria). These performances have made Expanscience the winner of the prize "Sustainable Beauty Awards" in 2013. This international award pays tribute to cosmetic companies acting in favor of sustainable development in the area of "sustainable packaging" for the Mustela brand.

[4] The Nagoya protocol for the access to and sharing of the advantages derived from the exploitation of genetic resources sets the principle of fair sharing of resources, their protection, and the obligation of consent of the countries concerned for the exploitation of the said resources and the acknowledgement of traditional know-how.

[5] The ISO 14001 norm is founded on the principle of continuous improvement of a company's environmental performance by limiting the impact of its activities on the environment. Since 2010, Expanscience has launched a system of environmental management and nominated a director for the ISO 14001 project. It has also established a leading committee to supervise the implementation of actions aiming at achieving a 20% reduction of energy consumption in the Epernon site between 2010 and 2015.

June 2014, our environmental performances showed a reduction of 22.4% in gas consumption, 26% in water consumption, and 16.5% in electricity consumption compared to 2010 levels.

Finally, our social responsibility policy toward our employees is a major axis of our CSR policy. As an actor in healthcare, we are evidently careful to the well-being of our personnel and the development of their skills in order to foster their professional progress and blooming. We have launched the "quality of working life" program, within which we have set up a week dedicated to the health of our employees in 2013. Since 2007, we have also carried out an innovative and participatory program, called "Graine d'ID" (seed of an idea). Its goal is to create an effect of synergy between different functions and individuals so as to better tap and develop creative potentials in every employee.

Logerais: As regards specificities of your activity, especially for dermocosmetics production, sustainable sourcing has been identified as the key stake. Can you elaborate on the fields of responsibility on this matter and how you manage it to reach tangible results?

Lemasson: Actually, the stakes are high. Beyond our responsibility as a global health-care company to our patients and customers, it concerns our responsibility to local people of countries supplying raw materials such as South Africa, Peru, or Mexico for instance. Here comes the question of our ability to ensure safety and durability of our supply through respect of local expertise and access conditions for plants, preservation of natural resources and environment, and also taking into account the local soil specificities, extraction conditions, and processing of products from natural ingredients.

Therefore, we have taken on a foundational job with each of our own plant supply chain in close collaboration with local partners, and conducted a detailed thinking about how to produce a "sustainable active cosmetic ingredient."

To understand how our actions are implemented, we can take the example of the avocado supply chain, one of Expanscience's strategic ones. Expanscience is the first global manufacturer of unsaponifiables of avocado and soy, and the Acacia supply chain.

Avocado is a key ingredient in many products marketed by Expanscience for the production of the medicine against osteoarthritis, in designing care products or active cosmetic ingredient. It is extracted in many forms such as oil, exfoliating powder, peptides, or sugars. The approach of Expanscience is to value the entire fruit at all stages of production: selection, slicing and drying, extraction, and final valorization.

As for the Acacia supply chain, it is at the center of a very aggressive policy of both biodiversity conservation and human development. It also illustrates perfectly the concept of "supply" or of "responsible sourcing."

In 2009, further to a merger with a local Burkinabe company specialized in ethnobotanical research, Expanscience identifies the Macrostachya Acacia forestry resource in Burkina Faso, from which we developed an active ingredient promoting skin hydration. This raises the question of exploitation and supply in order on the one hand to preserve local biodiversity and prevent desertification, and on the other hand to integrate the concept of economic and social benefit to local populations. Through the establishment of a tripartite partnership among the local community, a group of women pickers and Expanscience, these two criteria are translated into concrete actions: the organic certification of 50 ha of harvesting areas—amid threat of the Acacia resource due to local practices such as bushfires, overgrazing, droughts, or local heating requirements—creation of a nursery, training of 100 women for the manipulation of the natural resource, and financial support for pickers, especially by increasing access to micro-credit.[6]

Thus, we were able to combine innovation with local economic development, in line with our commitment to a "fair return" to partner countries in the framework of our supplies of vegetal raw materials.

Moreover, Expanscience has conducted a proactive approach of protecting traditional know-how and stemming bio-piracy. We call upon local experts who systematically gather and document expertise, for each and every plant supply chain, whether the knowledge exists in written or oral form. Expanscience then reports on this information in its scientific publications and in the introductory pages of its patents.

Our "sourcing" practices are now subject to regular assessment, as part of the company's commitment to transparency with UEBT via annual reports and audits conducted by a third party on our business.

Logerais: Through the examples you mention, you show how CSR may drive innovation as an answer to strategic stakes on your market.

Lemasson: Innovation is at the heart of our activity and the constraints we meet are driving us to think and design differently. For instance, within the context of our CSR policy, we have been working with our stakeholders and internal experts in 2010, in order to provide a precise definition, accessible on our website, of the principle of "naturalness" for the dermocosmetic brands of Laboratoires Expanscience. To date, this concept is not subject to harmonized regulation. This definition factors in most of the criteria, often restrictive, which allow us to select the active ingredients and raw materials making our formulations: the origin of the raw materials (for instance vegetal or mineral), the transformation processes carefully crafted to limit the impact on humans and environment to the minimum (processes approved according to the Ecocert referential are preferentially retained), the absence of questionable chemical substances such as paraben, phthalates, and phenoxyethanol. The absolute priority is to ensure the optimal safety of our formulations, without compromising another concept at the heart of our activity—ecodesign.

Research and development is not our only field of innovation. Combined to CSR, it essentially means integrating a new way of thinking and acting for the Expanscience workers in the way they execute their activities and to engage all the stakeholders in the corporate decisions.

---

[6] Expanscience has conducted a dozen activities for returning to Burkina Faso in partnership with the union of craftswomen "Ben Nafa Kabo, de Gassan." Besides advances on harvesting, Expanscience established in 2010 a cooperative microcredit for the acquisition of a parcel of land on which was built an office and a storage building for organic harvests.

Logerais: Do you think a CSR strategy can help in the challenging economic context of healthcare?

Lemasson: CSR is a process that involves and allows better visualization of flows and challenges working habits and production with an optimization goal. In an industry actually faced with the pressure of rising drug and price of resources, the economic issues related to sustainable development can be significant, e.g., for local plant supply chains in our partner countries. For example, in the case of the avocado oil production, the search, in partnership with our producer, for a new energy-saving electrical procedure yielded savings affecting the purchase price, hence allowing the producers to be more competitive.

Moreover, the savings from reductions in energy consumption in our own production sites can be used to make new investments.

Logerais: CSR is increasingly bringing the notion of "prevention" in healthcare. How do you deal with this trend? Do you feel comfortable to explain how it is compatible with your business model?

Lemasson: We believe we have an important role to play especially in the context of the emergence of medicine 3.0 where the patient and the general public are seeking greater autonomy in managing their health and well-being through access to high quality information. We approach the theme of prevention from two angles, by supplementing the drug treatment of diseases such as osteoarthritis by monitoring treatments and advices on healthy living through sports and nutrition to patients with these diseases or wishing to avoid them. For instance, we have launched a personalized coaching service for osteoarthritis sufferers (www.arthrocoach.com).

As a part of our CSR, Laboratoires Expansciences created the Healthy Advice and Prevention Pharmacies Club in 2012, which brings together pharmacy customers around prevention activities (osteoarthritis and sport, the mysteries of the baby's skin, etc.) to encourage the public to behave responsibly. By their mission and their proximity to customers, pharmacists have an important role in educating and mobilizing the public. Eventually, the civic role of the pharmacists is so important that they can make demands to pharmaceutical companies in terms of traceability and ecodesign of drugs and dermocosmetics.

While in France, no law requires the pharmacist or physician to prescribe medications whose molecules have a limited environmental impact; other countries, such as Sweden, have set up a classification of molecules according to a pollution index that is used to support the prescription, thus allowing better control of active substances ingested by humans and discharged in nature. Therefore, this is also a way to anticipate future potential evolutions.

Logerais: Your CSR Policy was recently evaluated by AFNOR and awarded an "exemplary" level of AFAQ 26000. What are the benefits of it, beyond image? Can you progress further?

Lemasson: Expanscience was recognized as "exemplary" by AFAQ 26000 in 2013 with a score of 708 points in 1000 and has now joined 4% of the organizations having reached this level at their first assessment, and is the only pharmaceutical and cosmetic laboratory.

These indicators contribute to our corporate image. They are very promising for managing risks related to our business, particularly in terms of reputation and access to raw materials, but the benefits of these performances are much deeper. They are a lever for creating shared value and support from employees whose sense of company belonging has increased, especially when the involvement of these staff members is an evaluation criterion for the performance of the approach, as it is welcomed by AFNOR.

The implementation of this appropriation by all 893 employees of Expanscience and its 12 subsidiaries remains today a priority objective with a high margin of progression to harmonize corporate practices across the Group, especially in responsible purchasing policy.

The CSR is increasingly integrated into the Expanscience governance mode through the growing use of its indicators in monitoring and management, the involvement of our stakeholders (environmental NGOs, patient associations, local authorities, suppliers of raw materials), and managerial innovation to link company's employees motivation to nonfinancial performance. Thus, in 2014, for the executives of Expanscience, the variable portion of their package is indexed partly (collective objective) on maintaining the "exemplary" level of the AFAQ 26000 assessment. This is a highly symbolic measure showing Expanscience's ambitions for the future.

About ISO 26000

In 2005, the International organization for standardization (ISO) wanted to lay down benchmarks for social responsibility in accord ance with international conventions in the areas of human rights, environment and labor regulations, complementary to the existing CSR initiatives Released in 2010, the ISO 26000 offers guidance for corporations and organizations on responsible social behaviors.

AFNOR Certification

Leading certification body in Europe, independent organization, AFNOR is pioneer in the evaluation of CSR approaches in France, with more than 200 private and public organizations granted AFAQ 26000. Its experts evaluate corporations then hand in an objective and unbiased review.

AFAQ 26000

AFAQ 26000 evaluates the degree of integration of sustainable development principles into corporations in alignment with ISO 26000 benchmarks and allows companies to demonstrate their transparency and maturity level of their CSR approach as well as their strengths and axes of improvement of their practices.

CPSIA information can be obtained at www.ICGtesting.com
Printed in the USA
LVOW01*1656191214

419643LV00001B/1/P